LISREL Issues, Debates, and Strategies

LISREL® Issues, Debates, and Strategies

Leslie A. Hayduk

The Johns Hopkins University Press
Baltimore and London

 Published 1996
Printed in the United States of America on acid-free paper
Typeset in Canada by LAH
05 04 03 02 01 00 99 98 97 96 5 4 3 2 1

The Johns Hopkins University Press, 2715 N. Charles Street, Baltimore, Maryland 21218
The Johns Hopkins Press Ltd., London

A catalog record of this book is available from the British Library.

Library of Congress Cataloging-in-Publication Data

Hayduk, Leslie Alec.
LISREL issues, debates, and strategies / Leslie A. Hayduk.
p. cm.
Includes bibliographical references (p. -) and index.
ISBN 0-8018-5336-2 (hc : alk. paper)
1. LISREL. 2. Path analysis—Data processing.
3. Social sciences—Statistical methods. I. Title.
QA278.3.H387 1996 95-44152
519.5'35—dc20(alk. paper) CIP

To Cherilyn, Daniel, and Vincent

Contents

Preface

Structural equation models are being employed in an expanding range of substantive areas, including:

addictions Embree and Whitehead, 1993;
aging Clarkson-Smith and Hartley, 1990;
counseling Mueller, Hutchins and Vogler, 1990;
criminology-delinquency Avakame, 1993; Hagan et al., 1989; Hayduk and Avakame, 1990/91; Kaplan and Johnson, 1991; Kleck, 1991; Matsueda, 1989, 1992; Norris and Kaniasty, 1992; Thornberry et al., 1994;
cultural studies Cui and van den Berg, 1991; Wiseman, Hammer and Nishida, 1989;
education Gustafsson and Balke, 1993; Reynolds and Walberg, 1991, 1992;
family studies Godwin, 1988;
health Beckie, 1994; Girodo, 1991; Hakkinen, 1991; Newcomb and Bentler, 1987; Sibrian and Elston, 1990;
human genetics Chipuer, Rovine and Plomin, 1990; Dolan, Molenaar and Boomsma, 1992; Hewitt et al., 1992; McArdle and Goldsmith, 1990; and *Behavior Genetics* Volume 19(1) 1989 is a special issue on LISREL;
intelligence Harnqvist et al., 1994; Jensen and Weng, 1994; Luo, Petrill and Thompson, 1994; Pedersen, Plomin and McClearn, 1994;
management Iverson and Roy, 1994;
marketing Lord, Lee and Sauer, 1994;
measurement Beland and Maheux, 1989; Benson and Bandalos, 1992; Byrne, 1989b; Saris, 1988; Thompson and Borrello, 1992;
medicine Hines et al., 1992; Horton and Horton, 1991;
methodology Acock, 1989; Alwin, 1989, 1992; Alwin and Krosnick, 1991; Batista-Foguet and Saris, 1992; Chan, 1991; Fleishman and Benson, 1987; O'Grady and Medoff, 1991; Schriesheim, Solomon and

Kopelman, 1989; Sikkel and Jelierse, 1987; Steyer and Schmitt, 1990;
nursing DeMaio-Esteves, 1990; Gulick, 1992; Johnson et al., 1993; Ratner et al., 1994;
organizational studies Lance, 1991;
planning Golob and Meurs, 1988;
political science Aish and Joreskog, 1990; De Graaf, Hagenaars and Luijkx, 1989; Hayduk et al., 1995;
psychiatry Romney, 1987;
psychology Breckler, 1984; Byrne, 1988; Church and Burke, 1994; Holahan and Moos, 1991; Keith, 1990; McCallum, 1990; Undheim and Gustafsson, 1987;
religious studies Stack, Wasserman and Kposowa, 1994;
robotics Offodile, Ugwu and Hayduk, 1993;
rural studies Armstrong and Schulman, 1990;
sexology Blanchard, 1994; Blanchard-Fields, Suhrer-Roussel and Hertzog, 1994;
social psychology Hayduk, 1994; Marsh, 1994; Marsh and Byrne, 1993; Owens, 1993; Rowe, Vazsonyi and Flannery, 1994; Sachs, 1992; Scanzoni and Godwin, 1990; Stryker and Serpe, 1994; Whitbeck et al., 1992;
socio-endocrinology Arnetz et al., 1991;
sociology Baer, Grabb, and Johnston, 1991; Botvin et al., 1993; Buhrich, Bailey and Martin, 1991; Hagan and Wheaton, 1993; Jenkins and Kposowa, 1990; Koslowsky et al., 1992; Rosenberg, Schooler and Schoenbach, 1989;
and ***speech/hearing studies*** Watson and Miller, 1993.

It is no longer sufficient to have and to present even a well fit model. One must stand behind one's model, and stake one's reputation on interpreting the model. Consequently, debates and disagreements over substantive models are beginning to be aired (Gillespie, 1991; Hays and White, 1987; Jagodzinski, Kuhnel and Schmidt, 1988, 1990; Marsh et al., 1994; Saris and Van Den Putte, 1988; Sheehan, 1987; Smith, 1989).

Another encouraging sign has been the appearance of critiques attacking the overall perspective of structural equation modeling, and responses defending this approach. See, for example, the multiple comments following Freedman (1987); the three chapters following Freedman (1991), namely Berk (1991), Blalock (1991), and Mason (1991); the November 1990 special issue of *Quality and Quantity*, and Turner (1987a,b).

LISREL was the first general program for estimating structural equation models, but it certainly is not the only program. LISREL's current competitors include:

AMOS (Arbuckle, 1988),
AUFIT (Browne, 1992),
CALIS (Hartmann, 1992),
COSAN (Fraser, 1988; Fraser and McDonald, 1988),
EQS (Bentler, 1985, 1992a)
EzPath (Steiger, 1989),
LINCS (Schoenberg and Arminger, 1988),
LISCOMP (Muthen, 1984, 1987),
LISREL (Joreskog and Sorbom, 1988, 1989, 1993),
MILS (Schoenberg, 1989),
Mx (Neale, 1993),
RAM (McArdle and McDonald, 1984),
SIMPLIS (Joreskog and Sorbom, 1993).

The mailing addresses for most of these programs are provided by Schoenberg (1989) or Waller (1993), and their alphabetical listing here implies that no endorsement of any of these programs is intended. Each program has its own peculiarities, strengths, and weaknesses. Waller (1993) provides a summary of the relative strengths of seven of these programs in the context of confirmatory factor analysis in the PC environment. EQS (BMDP based), CALIS (SAS based), and LISCOMP are being used with increasing frequency but LISREL is still the most widely used structural equation program. I use LISREL in this book because of its popularity, because its notation has been widely adopted, and because it is being continually updated and improved by Karl Joreskog and Dag Sorbom.

There was a big change in the LISREL manuals between LISREL 7 and LISREL 8. The LISREL 8 manual introduces SIMPLIS, which is a version of LISREL written without matrix algebra or Greek characters and which is explicitly designed "for students and researchers with limited mathematical and statistical training" (Joreskog and Sorbom 1993:i). This permits LISREL to compete with EQS and AMOS which had an edge in novice-friendliness, though it means that serious users will also want a copy of the LISREL 7 manual to which Joreskog and Sorbom repeatedly refer the reader for technical details.

The absence of a general notation in SIMPLIS will seem efficient to infrequent users since they need not learn any notation, but there is a price to pay. The absence of a notation system means that there is no convenient way for SIMPLIS users to represent the standard segments of models. SIMPLIS users will be hampered in their attempts to inform their readers about which of their variables are latent concepts and which are indicators, which concepts are exogenous and which are endogenous, or to discuss similar features from multiple models (such as the sets of LISREL λ's from two models). These standard demarcations are entrenched in LISREL's notation precisely because they are useful and assist in the discussion of models. If SIMPLIS users wish to organize the discussion of their models in any way other than merely listing equation after equation, they will either have to learn LISREL notation, or struggle with long lists of variable names as they repeatedly regenerate the routine demarcations in each model they ever run.

Most of the LISREL 8 manual is dedicated to redoing the basic examples from the LISREL 7 manual in the new setup style, but there are a few new features. Most notably, LISREL 8 permits nonlinear constraints which should aid in the estimation of interactions (Jaccard and Wan, 1995), and it introduces a theta-delta-epsilon ($\theta_{\delta\epsilon}$) matrix which permits easier specification of error covariances between the indicators of the exogenous and endogenous concepts. TH(1,3), for example, refers to the covariance between the errors on the first exogenous indicator and the third endogenous indicator. Loadings crossing between the exogenous concepts and endogenous indicators (or vice versa) still require an all-η model specification. Given LISREL 8's slant towards those with limited statistical training, it is a bit incongruous to find that LISREL 8 also introduces the technicalities of a host of new fit indices.

But my intent is not to repeat or describe the latest features of LISREL. Instead, this book takes a much broader view of the structural equation modeling exercise, and attempts to narrow the gap between the models we employ and the research questions relevant to advancing specific research domains. How is it that we can best employ structural equation models to maximize the likelihood of making substantive research advances?

To adopt such an approach, I have had to assume that the reader has a basic understanding of structural equation modeling and LISREL notation. When a reference to a basic discussion is required I refer the reader to *Structural Equation Modeling With LISREL: Essentials and Advances* (Hayduk, 1987), hereafter simply *Essentials*, though I realize the reader may have learned structural equation modeling through other sources (e.g., Aaronson, Frey and Boyd, 1988; Alwin, 1988; Bentler, 1988; Bollen, 1989a; Boyd, Frey and Aaronson, 1988; Byrne, 1989a, 1994; Cole, 1987; Cuttance and Ecob, 1987; Fassinger, 1987; Lavee, 1988; Loehlin, 1987; Lomax, 1989; Mulaik, 1988; Nunnally and Bernstein, 1994; Pfeifer and Schmidt, 1987; Swaminathan, 1991; and Von Eye and Clogg, 1994). I have tried to assist readers from these other backgrounds by providing summary descriptions of each "point at issue," so those unfamiliar with *Essentials* can locate the corresponding material in the texts with which they are familiar. On the topics of loops and phantom variables, *Essentials* is the most comprehensive available source, but on most other topics one can find practical advice from any of several sources (e.g., Anderson and Gerbing, 1988; Bentler and Chou, 1987; Bollen and Lennox, 1991; Breckler, 1990; Cliff, 1983; MacCallum, 1986; and Steiger, 1988).

I began this book by considering what it was that was keeping researchers currently using structural equation models from making important advances through the use of LISREL. Having located a variety of topics that I felt warranted addressing, I eliminated those topics that I felt others were likely to address (Hoyle, forthcoming), and focused on those topics where I thought I could provide some assistance. Sometimes the assistance took the form of reviewing the overall structure of the enterprise (Chapters 1, 4, and 6). At others times it meant summarizing a debate so that we could come to a fair adjudication of the outcome (Chapter 2 on the two-step procedure). It also meant enquiring into areas where no one else was likely to venture (Chapter 3 on implications of loops, and Chapter 5 on stacked models based on different variable sets). As I struggled with these topics, several common themes emerged. Most notable among these are concerns for equivalent models, researchers' ownership of their models, and the blinders set in place by the factor model.

Before I turn to these themes in overviewing the chapters, a few general points need to be aired. First, the approach I have taken is not to examine the details of statistical estimation, but to examine the fundamentals of the use, application, and interpretation of models. I am most interested in investigating how structural equation models can prod us to think differently, and how they can assist us in the research enterprise. I hope this emphasis on "why we are doing what we are doing" lends a flavor to this work that parallels the irreverence that Austin and Wolfle (1991:120) detected in *Essentials*. It is not that I wish to be irreverent; it is merely that there are many issues where researchers should not look to statisticians for the answers. Statisticians simply do not see the issues from the same perspective as someone who cares about the substantive research domain. This is not to fault the statisticians. It merely acknowledges that they have a different stake in the exercise; a stake which they cannot undo by merely adding another subroutine. I hope the differences between the statisticians' and researchers' perspectives become obvious as we proceed, though I have tried to avoid antagonistic cross-characterizations, as no offense is intended. Both perspectives have their role to play.

I must also explain the figure that follows. This figure is a pedagogic device I have found useful for summarizing the foundations of structural equation modeling. We will have occasion to refer to this figure as we proceed. The top row of the figure represents the model we care about, whether that model is expressed in words, as a figure, or as equations. The model, in any of its three equivalent forms, is there by postulation, and constitutes a statement that "If the world works this way..." The connections between the verbal, diagrammatic and equation forms of this "if" rest on grounds that statisticians are unlikely to defend or enquire about, yet this is precisely where the figures and verbal descriptions that populate the researchers' world become attached to the equations that populate the statisticians' world.

The Implications of Structural Equation Models

		"X causes Y"	"X1 and X2 both cause Y"	"X1 causes both X2 and X3"	"...these influence these..."
IF a model		X —b→ Y 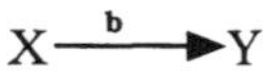 $Y = a + bX$ eq. at top of page 14	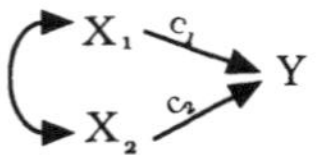 $X_1 \to Y$ (c_1), $X_2 \to Y$ (c_2), $X_1 \leftrightarrow X_2$ $Y = c_1X_1 + c_2X_2$ eq. 1.24 with only two X's	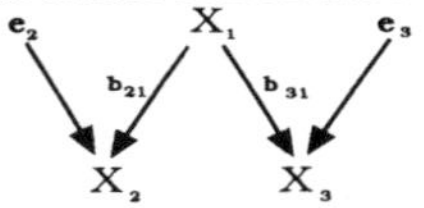 $e_2 \to X_2$, $X_1 \to X_2$ (b_{21}), $X_1 \to X_3$ (b_{31}), $e_3 \to X_3$ $X_2 = b_{21}X_1 + e_2$ $X_3 = b_{31}X_1 + e_3$ eq.'s 1.55, 1.56	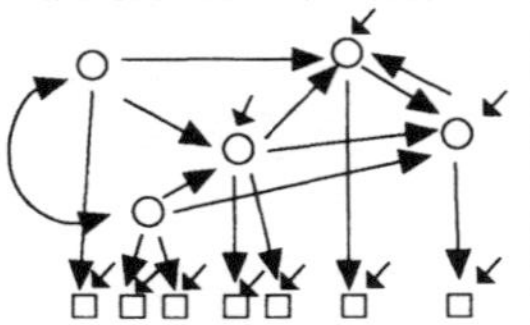 eq.'s inside front cover or on page 91
T H E N	means or expected values	$E(Y) = a + bE(X)$ $\overline{Y} = a + b\overline{X}$ eq. 1.16	$E(Y) = c_1E(X_1) + c_2E(X_2)$ eq. 1.25 with only two X's	SPECIAL CASES OF: eq.'s 1.58, 1.59	Section 9.2 pages 286-322
	variances and covariances	$Var(Y) = b^2Var(X)$ eq. 1.23 $Cov(XY) = bVar(X)$ eq. at top of page 17	$Var(Y) = c_1^2Var(X_1)$ $+c_2^2Var(X_2)$ $+2c_1c_2Cov(X_1X_2)$ eq. 1.28 or eq. 1.26 with only two X's	VARIANCES OF X1 and X2 WERE NOT GIVEN BUT ARE SPECIAL CASES OF: $Cov(X_2X_3) = b_{21}b_{31}Var(X_1)$ eq. 1.67	eq. 4.62 page 116
	higher moments				

Note: Equation and page numbers refer to Hayduk, 1987. *Structural Equation Modeling with LISREL: Essentials and Advances.*

The lower rows of the figure provide the "then" or observable characteristics of the data that must arise as a consequence of postulating the model heading the column. Specific connections among the model's structural coefficients and the observed means, variances, and covariances are demanded for any particular model. In this way, the theories embodied in models make inescapable predictions, and it is the veracity of these predictions that provides the glow of success or the sting of failure.

The first three columns of this figure constitute an outline of the implications of the models considered in *Essentials* Chapter 1 (located by equation numbers referring to *Essentials*). At several points we draw upon the implications of these basic styles of model components. If you have not read *Essentials*, you should review these columns to become comfortable with the unavoidable implications of single, multiple, and common-cause model components, and to locate where in your personal reference base you can turn to review these claims.

The final figure column draws the parallel between any particular full LISREL model and the strict and rigid implications that model has for covariances and means. We need to be very clear that every model has specific and undeniable consequences, and that the undeniability of the consequences comes from the mathematical proof that links the equations representing the model to the covariance matrix of the observed variables. This is the proof that links the LISREL manual's equations 1.1, 1.2 and 1.3 to equation 1.4 (Joreskog and Sorbom, 1989), as discussed in *Essentials* (106-116). For any given model, the consequences are the specific connections between the structural coefficients that comprise the model and the properties of the joint distributions of the observed variables in the model, though the details of these connections are too extensive to compress into the figure.

These consequence are causal consequences if the slopes in the model's structural equations are interpreted as causal effects. Note that causation is an attachment to the postulated "if" and is not necessarily in the real world. Causation is a part of our comfortable ways of thinking, guessing, postulating, or making an "if" about the world. Given that statisticians have no objection in principle to interpreting slopes and directed arrows in models as causal effects, the only real issue about causation is whether the world is similarly causally segmented and directed. The discussions of equivalent models that follow can be thought of as differing causal segmentations which have identical implications for the observed covariance data, and hence which are indistinguishable from the perspective of the available covariance information.

The "higher moments" referred to in the last row of the figure allude to the fact that models also have implications for properties of distributions other than the means, variances, and covariances of the variables. For example, the skew and kurtosis of causally prior variables contribute to the skew and kurtosis of causally subsequent variables. This row has been left blank because we will not be addressing this particular issue. Most of the following discussions focus on variances and covariances, though there is nothing which in principle halts the extension of the ideas to cover means and higher moments.

The redundancy of the figures and equations in the "if" row of the figure is exploited by Joreskog and Sorbom (1993) to permit LISREL 8 to translate figure modifications into corresponding structural equation modifications. But the redundancy of words, figures, and equations as model representations deserves a more fundamental place in our thinking. Interpreting a model depends on the redundancy between the equations and our verbalizations. And developing a model usually amounts to starting with the words and ending with the corresponding figure and equations. Styles of verbalization that cannot be translated into equations point to the current limits on structural equation modeling. Indeed, *every style of verbalization that cannot be converted into a corresponding mathematical model constitutes a fundamental boundary on the range of applicability of structural equation modeling*. And every instance in which the untranslatability of a particular segment of a verbal model has led to the exclusion of that segment from the statistical model challenges the adequacy of the resulting model. If the verbal to equation translation is incomplete or imprecise, the researcher must admit to being unable to specify the model that was really wanted. Hence, the researcher must also concede to having estimated, at best, a nearly correct model, or at worst, the wrong model.

Fortunately, many styles of verbalization have proven to be representable as structural equation models, but this should not deter us from continually attempting to demarcate the verbalization styles that currently cannot be translated. Each style of verbalization that cannot be translated into structural equations locates a class of theorizing that is in principle unmodelable by LISREL. Threshold, conditional, or buffered effects (as when varying resources differentially offset some burden) come to mind. Uncertainty or imprecision in the translatability between a model's verbal and mathematical representations leaves one unsure as to whether a researcher actually tested the theory he or she intended to test, or whether a particular interpretation can be justifiably attached to a specific model. Any looseness in moving in either of these directions contributes to confusion, not to freedom or flexibility.

At times, an inability to locate a corresponding mathematical model can be overcome by adopting new or creative modeling strategies, but we should remain alert to the fact that some particular verbalization styles simply cannot be made consistent with any set of equations. We should note that it is doubly wrong, though not all that uncommon, for researchers who have been unable to specify their desired model to estimate a nearly correct model and then to provide a compensating misinterpretation of that model.

Statisticians are unlikely to be of much help in making decisions about what differences between one's words and equations are sufficiently fundamental to halt a modeling attempt. Disciplinary jargon is always a touch loose, and statisticians cannot be experts in all the substantive areas to which structural equation models are applied, so the statistician slips off the hook. The warning is clear: let the researcher beware. Your career hangs on the precision with which you can meld your discipline's words to your structural equations. Articles reporting on the difficulties encountered in translating disciplinary jargon into a model would sharpen our understanding of both the disciplinary jargon and what current structural models are incapable of investigating.

An appreciation of the "if" in the preface figure also weans one off a dedication to real models, and replaces this with a dedication to realistic models. Realistic models, for example, might require the insertion of reasonable but unmeasured segments into models (Hayduk and Avakame, 1990/91) or the estimation of effects from a possible yet currently unidentified variable (such as the BACKGROUND variable used by Matsueda, 1989).

Another general orientation that pervades the following chapters can be summarized in the phrase "model all your methodological problems." Methodological problems are things that have undesirable influences on the data. The undesirability of these things often lead us to over look the fact that these are influences or effects. As effects, these are often representable as specific model components. Even if one is unable to estimate the relevant kind of component, due to the absence of appropriate measured variables, one can usually introduce model components that postulate, and hence adjust for, interfering effects of approximately the proper size (Hayduk and Avakame, 1990/91). By incorporating the methodological problem as a part of the model, the researcher obtains estimates for the remaining substantive parts of the model which have adjusted, compensated, or controlled for the unavoidable methodological problem.

Essentials allocated relatively few pages to specifically discussing factor analysis and models with multiple indicators. In contrast, factor analysis and multiple indicators are extensively discussed in Chapters 1 and 2 below. This emphasis arises because the factor and multiple indicator models highlight many of the common threads underpinning a variety of points that warrant debate and comment. At several points I will criticize the factor model, but no blanket condemnation of the factor model is intended. The factor model is simply another style of model, and it may or may not be appropriate for any given data set. Problems arise only when researchers become so wedded to the factor model that it seems like it is the only model.

Let us switch our attention now to the more specific contents of the chapters. Chapter 1 begins with a consideration of the "if" from the preface figure, in the context of measurement. Specifically, it defends the position that we should not be apologetic about the fictitious, or iffy, nature of models, including the measurement portions of our models. This leads to a discussion of how we can recover some solidity to measurement despite the fluidity introduced by admitting that the measurement segments of models are postulated if's.

The solidity is found in the notion of model constraints, which I address in the context of multiple indicators, common causes, and the factor model. In addition to demonstrating how factor models gradually shade off into nonfactor models, we consider the necessarily proximal nature of factors in Section 1.4 and end the chapter by reviewing how multiple indicators should be incorporated into structural equation models. The procedure I recommend maximizes the researcher's control over the meanings of the latent concepts and provides an integrated procedure for the modeling of single and multiple indicators. A single indicator is modeled using the same procedure that is used to model the first indicator of a set of multiple indicators.

The discussion of measurement and the factor model in Chapter 1 leads directly into the Chapter 2 discussion of whether a factor model should routinely be estimated as a first step, prior to estimation of any second-step structural model. This two-step procedure has been proposed by Anderson and Gerbing (1988) and Joreskog and Sorbom (1993), among others, but it has also been criticized (Fornell and Yi, 1992a). Chapter 2 reviews the various claims and then extends the debate into discussions of the use of scales as measures (which I view as inadvisable), and the dimensionality and meanings of concepts. I conclude that the two-step procedure is not likely to be particularly helpful, though the more important observations return us to questions of measurement, the iffy nature of models, researchers' assertiveness about their conceptual meanings, and the possibility of modeling concepts with no indicators. The changing target for χ^2's probability, as reported in the point form summary in the final section of this chapter, will be of interest to most researchers, not just those considering the two-step procedure.

Chapter 3 shifts away from measurement and the factor model but continues the focus on the iffyness of models. This chapter highlights the importance of locating and discussing equivalent models, that is, of locating and discussing if's which have equivalent then's, implications, or consequences. This chapter introduces a previously unrecognized class of equivalent models, namely models inserting a loop into an otherwise recursive model. One of the implications arising from this discussion is that all the estimates in all recursive models can be viewed as biased if they are viewed from the perspective of the equivalent loop model. It also turns out that loop models provide a very natural interpretation style for longitudinal models, and for models with replicate measures.

Among this chapter's many twists and turns we encounter a perfectly acceptable and interpretable model containing a negative squared multiple correlation. I used to think that no acceptable model could contain a concept having a negative R^2, but I was wrong! Some implications of the material in this chapter are devastatingly clear, but overall this chapter is sufficiently novel that I do not feel I have personally fathomed, let alone presented, all the implications of this material. This chapter provides considerable food for thought.

Chapter 4 examines the topic of equivalent models from a different perspective. It begins with a summary of equivalent models provided by Stelzl (1986) and then places this in the context of the logic underlying the TETRAD program (Glymour et al., 1987). LISREL and TETRAD are in some ways complementary programs, and in other ways they are competing programs. On the competition side, I consider how it is that TETRAD was able to outperform both LISREL and EQS when all three programs were asked to suggest model modifications that would transform an incorrect initial model into the true model. On the complementary side, TETRAD can be used to locate data-prompted revisions to a model while LISREL provides the estimates of the coefficients in that revised model.

The point of this chapter is not to push you into learning TETRAD, but to introduce some of TETRAD's thoughtways into our LISREL modeling habits. One should be able to do a better job of model modification and model development if one pays attention to some general operating principles that are consistent with TETRAD. I anticipate that TETRAD-like capabilities will eventually be appended to LISREL, so this chapter also serves as an anticipatory explanation.

Chapter 5 turns to a discussion of what is to be gained by considering stacked models based on differing numbers of indicator variables. The possibilities include identifying otherwise unidentified models, controlling for unavailable variables, and integrating unmatched data sets. LISREL does not formally permit stacked models based on different sets of indicator variables, but this can be accomplished using some tricks of the trade. This chapter will be of interest to anyone whose data set includes variables measured for only some of the sampled cases. If question-routing steered some of your respondents around a specific set of questions, or if an indicator was too expensive to collect for all your cases, this chapter is for you.

Chapter 6 is intended to tie up some loose ends. First it examines the coefficient of determination that has been reported from LISREL V onward. I suggest that this coefficient be abandoned as useless. The next section of the chapter challenges statisticians doing Monte Carlo studies to develop a more realistic test model than the traditional factor model. I propose a motif model that encapsulates a more typical and important set of model characteristics. This motif model is complemented by a nuance model that incorporates a set of interesting but less common features.

The chapter concludes with a survey of the recent literature. This includes some discussion of some fit indices and the expected parameter change statistic, but my primary purpose here is to report where the rest of the literature has been going in the years since I wrote *Essentials*. It is my hope that this section speeds your search for recent material on whatever other structural equation topics are of concern to you. My relegation of topics like polychoric correlations, multilevel models, and other technical statistical issues to brief summary reviews is not a sign that I view these as unimportant. It merely says that I do not believe that I currently have enough to contribute on these topics to make it worth your while reading me rather than the literature.

Having drawn the boundaries on the playing field, it is time for the academic games to begin!

Acknowledgments

I thank Pamela Ratner, Joy Johnson, Joan Bottorff and Rainer Stratkotter for their detailed comments on the drafts of the chapters that eventually became this book. Their efforts have improved the readability, cogency, and clarity of numerous sections of this work. I also thank Mike Gillespie, Guy Germain, Bill Sveinson, Dennis Bray, Frank Grigel and the 1995 LISREL group for their comments and assistance. Clark Glymour generously provided me a draft copy of the TETRAD II users manual which assisted in the preparation of some of the material in Chapter 4. I also gratefully acknowledge the computer and other support of this book provided by the Department of Sociology of the University of Alberta.

Chapter 1

The Constraints and Deceptions of Factor Models

LISREL began as one of a set of programs Karl Joreskog developed at Educational Testing Services in the 1960s. The most widely known of LISREL's sister programs were EFAP (Exploratory Factor Analysis) and COFAM (Confirmatory Factor Analysis), both of which are relatively infrequently cited today. COFAM was a submodel of the basic LISREL model and was absorbed into LISREL. This fostered a group of LISREL users who do factor analysis within the confines of the LISREL program. Barbara Byrne's (1989a) book *A Primer of LISREL*, for example, is entirely devoted to factor models.

The factor model is as good as any other model but problems arise when researchers become so wedded to a particular model that they fail to appreciate the alternatives, or to recognize fundamental inconsistencies between their model and their own thinking. This chapter examines how model-myopia can lead to overlooking useful data, and to engaging in "traditional" research practices that hide potential problems. Unfortunately, many of the afflicted individuals do not recognize their implicit dependence on the factor model, preferring instead to speak of "a scale of X," or "measures of Y."

This chapter ostensibly focuses on the factor model, but the issues we raise are intended to address measurement issues more generally. We begin by discussing models as fictions to which we should be committed. Measurement is part of that fiction. We then develop an appreciation for what this commitment commits one to, by investigating the simultaneous flexibility and rigidity of models. We do this via an extended discussion of the proportionality constraints inherent in factor and multiple indicator models. This provides an opportunity to clarify and extend the procedures for handling multiple indicators that were presented, but not headlined, in *Essentials*.

In addition, this discussion provides a foundation for our investigation of the two-step procedure in Chapter 2, and model modification, equivalent models, and TETRAD in Chapter 4. The two-step approach proposes the routine development of a measurement model (a step-one factor model) prior to considering a structural (step-two) model, and hence highlights issues implicit in the factor analytic tradition. Model modification, equivalent models, and TETRAD demand a firm grasp of the idea of model constraints. The following assumes the reader has encountered at least some minimal prior discussion of model constraints, so I am free to provide a slightly different slant on this fundamental component of structural equation thinking.[1]

1.1 Models as Fictions We Should be Committed To

Models are unrepentant and unapologetic fictions. We create models as accompaniments to, and ultimately replacements for, the brand of storytelling called theorizing. The value of a model, however, does not arise from the fanciful nature of its concepts or the thrust of its effects. A model's primary value arises from the constraints imbedded in that model. As labile and fictitious as models are, they are in other ways unbending, unforgiving, and rigid.

The unyielding constraints implicit in models confront researchers in two ways. A visible confrontation occurs when the rigid placement of specific null effects within a model makes that model fail. Despite the freedom to make some model modifications, and to select coefficient estimates that minimize the discrepancies between the data and the model's implications, there are some models whose χ^2's remain stubbornly significant.[2] A bad model (sore thumb) sticks out precisely because it is rigid. Without such rigidity there would be no alternative or competing models. It is only when a model stands solidly against some specific connections between variables that the model can be evaluated in comparison to different, but equally rigid and uncompromising, models.

A less visible, but more important, encounter with a model's unyielding rigidity may occur in trying to develop a model that encapsulates one's theory. The issue here is: do the model's constraints match the theory's constraints? Any mismatch here is devastating because it implies one is testing a theory (model) other than the theory one claims to be testing. Mismatched constraints of this type are much less likely to be noticed because they are never challenged by any program output. They are only found when reviewers, or other antagonists, note that the model is inconsistent with the author's interpretations. And we know how slippery some authors' interpretations are!

Models, then, are imbued with two seemingly contradictory features. They simultaneously exhibit the boundless flexibility of fantasies and the inflexibility of rigid constraints. The fantasy speaks to the creative and imaginative side of science, the constraints are the fertile soil from which one derives specific consequences as *necessarily* following from one's particular fantasy.

Much of the art of good LISREL modeling is in attaining a close match between the verbal fictions we call theories and the mathematical fictions we call models. The quality of the match is determined by the precision with which the rigid components of the model correspond to the uncompromising aspects of our verbalizations. If one's model is rigid about a feature the theory treats as loose, or if one's model is flexible where the theory is adamant, one has not yet mastered the art of model construction.

It is relatively easy to get researchers to make parallel theoretical and model claims about the fictions called "effects." It is more challenging to gain commitment to the "absence of effects."[3] One must be able to provide an affirmative response to the question: "Have I included all the effects that my theory stands for?" This self-query is usually followed by a pause, as the researcher is repulsed by the possibility of letting "everything influence everything else" in response to the competing claims in the literature. Any hesitancy in accepting the assertion that there is no need for further effects is a sign of problems. Without some substantial commitment to this, the modeling exercise is not yet worth starting.

My advice is to make a model reflect not the literature but only one identifiable position from the literature. Each position or perspective should be granted a separate model, and hence models in general should not be complex composites of competing conceptual views. Structural equation models provide a greater service if they assist us in clarifying the implications of specific theoretical positions than if they precisely test poorly demarcated theoretical positions.

Good modeling requires both the imagination to create the fiction and the courage to display a commitment to that particular fiction. Any theoretician persistently unwilling to make such a commitment will be viewed as lacking dedication to any short-list of features, and will suffer the consequences of perpetual academic obtuseness. These theoreticians are doomed to mere verbal fictionalizing, where there is minimal hope of finding, let alone capitalizing on, any implication structures.

By commitment I do not mean one should defend a model despite disconfirming evidence. The commitment is to maintain the integrity of the theory/model, or the style of argumentation that constitutes the verbal foundations of the model. This is a difficult challenge because we are usually taught to synthesize, combine, or meld perspectives rather than to stand each perspective clearly in opposition to all others, and therefore to attend to the characteristics that define or bound perspectives. The precision displayed in drawing and maintaining (being committed to) these boundaries lifts one's theoretical/verbal precision a notch or three, and provides a contribution that is as valuable to a discipline as are any estimates or model tests.

Most sociological LISREL models are created without guidance from theoreticians. The theoreticians may feel shunned, but this seems to have no ill effects on the models. I take this as a positive sign because it indicates a bountiful store of ingenuity wedded to a commitment to delineating, preserving, and ultimately testing recognizable theoretical positions. What more can a research area ask of its practitioners?

A theoretician/researcher who is not committed to the story being told by a model will disown that model when it comes to the number-crunch called a significant model χ^2, and little will have been accomplished. This is not to suggest that researchers with a highly significant χ^2 should bow their heads in shame for having failed to locate the correct model. I would prefer to see many more published discussions of failing models (see for example Hayduk and Avakame, 1990/91). Public discussion of failing models raises the precision displayed in interpreting models and, hopefully, should assist others in avoiding similar theoretical pitfalls.

Science can function admirably with substantial doses of random foraging, such as inheriting a discipline's traditional models, but this works only when it is tempered with publication of both successful and unsuccessful models. Reviewers and editors who routinely discourage publication of failing models are harming, not assisting, their research communities. But to me the real culprit is the researcher who abandons the failing model. The researcher is the only person who can provide the precise, critical, and cautionary manuscript. The difficulty seems to be in getting researchers to be sufficiently honest with themselves to objectively review their own model's failures. Research associates can provide a great service to their disciplines by keeping the "I think we should change models" rug from being pulled out from under failing models. If a model is worth estimating, it is worth reporting, whether or not it fits the data.

One reward for having committed oneself to a match between a particular model and one's theory is that one can draw upon the implication power of mathematics. This is usually glossed over as a mere step along the way to the estimation of model coefficients. The maximum likelihood estimates, for example, are merely those coefficient values that minimize the discrepancy between a model's rigid and undeniable implications (recall Σ as a model implied covariance matrix)[4] and the equally inflexible data (the covariance matrix S). That is, estimates are obtainable only when one is stuck between a rock-solid commitment to a model's implications and a hard-data place.

It is not merely the ability to obtain estimates, but the fact that a model has specific identifiable and unavoidable implications that is newsworthy. In *Essentials* I presented the basic LISREL model setup and immediately followed this by the "difficult" mathematics of how a model has implications (106-116) to highlight how specific models imply specific consequences. It is only when one sees a structural equation model as more than a bunch of equations that one can appreciate how a model can be a tool for investigating the consequences of a way of speaking. An appreciation for the way those consequences are arrived at permits a deeper understanding of what a model is about, and hence permits a clearer conception of how one might adapt one's discussion of a particular structural equation model to coincide with the needs of one's discipline. A discussion of how a particular coefficient contributes to a particular observed covariance (e.g., by helping to account for it spuriously) is often more informative than a report of the precise magnitude of that coefficient's estimate.

1.2 Measurement and Structural Effects as Fictions

Most researchers admit the fictitious nature of "effects" (β's and γ's) as soon as "effects" are rephrased as "causal effects." This seems to evoke recollections of the public contortions and gyrations of generations of philosophers. These same researchers, however, usually resist the notion that all measurement of concepts is equally fictitious. Measurement seems so solid and indispensable that this couldn't possibly be part of a fantasy world. Surely, measurement is precisely what brings our theoretical fantasies down to earthly observables. The grounding of fantasies in recorded facts must be more than mere imaginings!

Admittedly, the β and γ effects in models link two fictitious concepts (latent variables), so one is seemingly at a distinct disadvantage if one attempts to remove these from the realm of fantasy. But does the presence of the observed variable at the arrowhead end of each λ in LISREL's measurement equations assist in reducing the hypothetical nature of these effect coefficients? I think not. I view these as equally fictitious.

Suppose I am required to justify my belief that concept η_1 influences y_1, that is to justify entering $\lambda_{1,1}$ as a coefficient to be estimated in my model, as depicted in Figure 1.1A. All the acceptable styles of justification amount to postulating some intervening connections between the concept and the indicator.[5] I might postulate one or more intervening variables as in Figure 1.1B, or I might concoct a more complicated story for how η_1 makes itself apparent in y_1, as in Figure 1.1C, but whatever story I come up with, it is fiction. It postulates new concepts, and hence returns us to the admittedly

Figure 1.1 Justifying a λ.

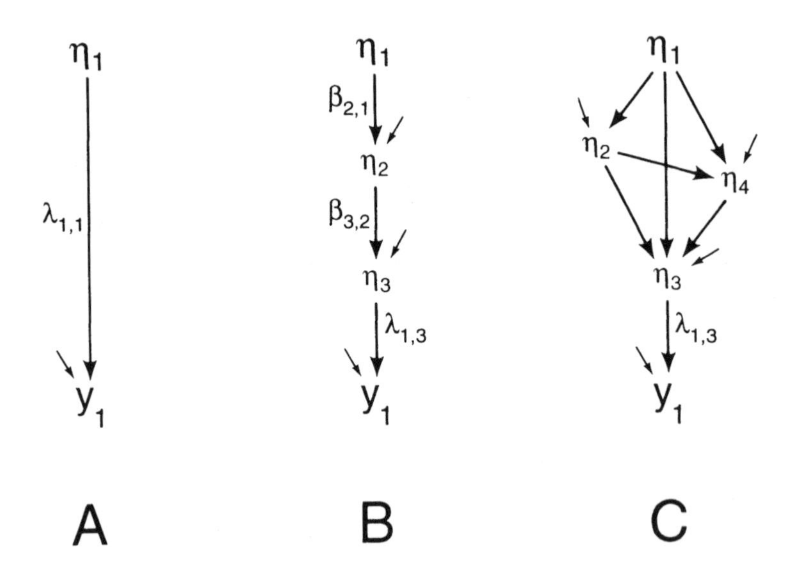

fictional world of β and γ effects, which are antecedent to a new "final arrow" whose head ultimately touches the "real" y_1 variable. That is, the original source of concern (the $\lambda_{1,1}$ in Figure 1.1A) has vanished, and has been replaced by some fictionalizing and a "new" $\lambda_{1,3}$ that will prove to be equally elusive by vanishing as soon as we focus our analytical attention on it.

I cannot justify $\lambda_{1,1}$ by appealing to the obvious and known nature of y_1 alone, because it is precisely my conceptualization of η_1 that is under attack when I am pressed to justify how η_1 connects to y_1. While all estimated λ values are in principle challengeable in this way, standard operating policy within each discipline sets practical limits on reviewers and critics by making them seem too picky or philosophical if they incorrigibly demand such justifications.

If we grant that measurement effects, λ's, are as fictional as effects among abstract concepts, have we not robbed all LISREL models of their solidity and testability? No, we have not. The χ^2 test of model fit is still there, and the data are still capable of rejecting our model/fiction. The key point is that the solidity of models is not to be found in the measurement or structural effects that comprise the model, or even in the true identity of the abstract (fictional) concepts in the model. It is to be found elsewhere, namely in the *constraints implied by the model.*

My task in the following section is to assist you in developing a sense of the rigidity of models that persists despite the boundless flexibility provided by the fictional concepts and effects comprising our models. To do this, we must examine a particular model. I chose the factor model because it is commonly used, frequently misunderstood, and occasionally worshipped. While the topic is the factor model, our aim is to foster a familiarity with model constraints in general.

1.3 Proportionality Constraints in Factor and Multiple Indicator Models

Let us begin by previewing the overall structure of Figure 1.2. The model in Figure 1.2A depicts five indicators as arising from a common cause, η_1, and may be equivalently described as a factor model with one common factor, or as multiple indicators of a latent concept. Figure 1.2B sidles imperceptibly away from ordinary factor analysis by asking us to imagine that one of the original indicators, y_5, really measures a different concept, η_2, which is correlated with the original concept η_1. While we might describe this as a "two-factor model" with correlated (oblique) factors, this is not a factor analysis that would be encountered by anyone using traditional factor analytic procedures. Traditional factor analysis programs cannot function with factors having only single indicators, and they often have no provision for implementing the fixed 1.0 loading, or λ, we have specified as linking η_2 to y_5.

Figures 1.2C, D and E move even further away from traditional factor analysis by allowing directed effects between the concepts/factors. A portion of each of these models looks like a factor model (η_1 being the factor common to items y_1 to y_4), but traditional factor analysis procedures do not permit the estimation of directed effects among the factors/concepts.

Overall, the subsections of Figure 1.2 progressively take us from a model that is entirely factor analytic in its thinking, to models that bear a strong conceptual and statistical resemblance to factor models, yet which are more appropriately described as structural equation models incorporating multiple indicators of one of the concepts, factors, or latent variables. These models are designed to highlight the implications of postulating multiple indicators for concepts by drawing a fundamental parallel between the factor model and multiple indicator models.

Figure 1.2 Proportionality in Factor and Multiple Indicator Models.

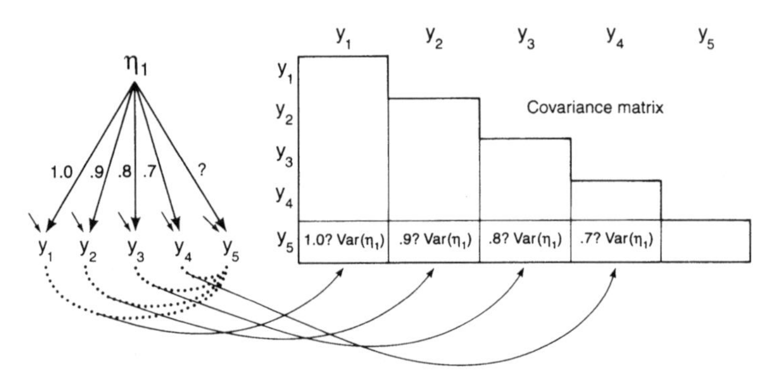

A: Proportionality within a set of indicators.

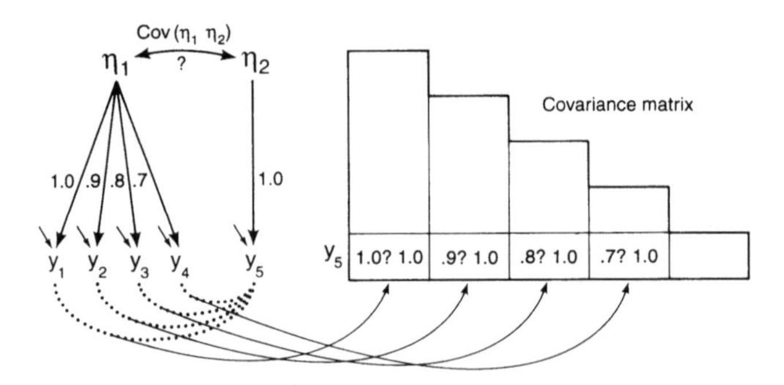

B: Proportionality between indicators of correlated concepts.

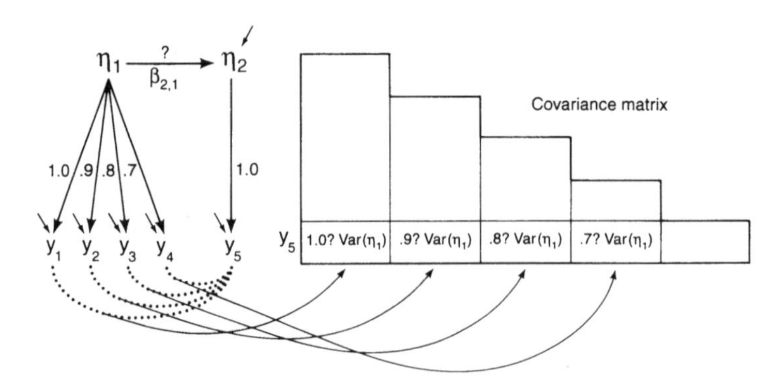

C: Proportionality to an indicator of an effect.

Figure 1.2 Continued.

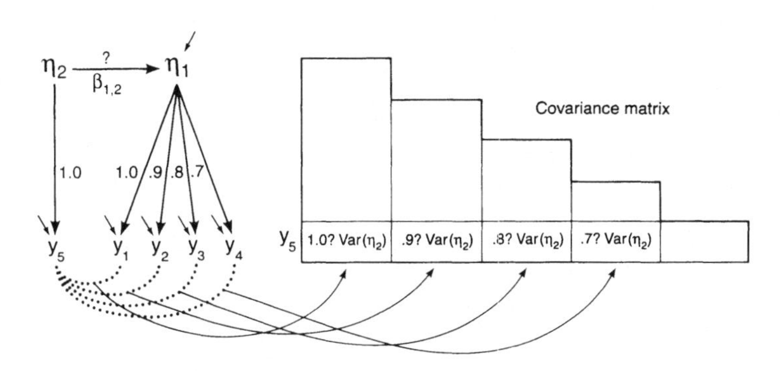

D: Proportionality to an indicator of a cause.

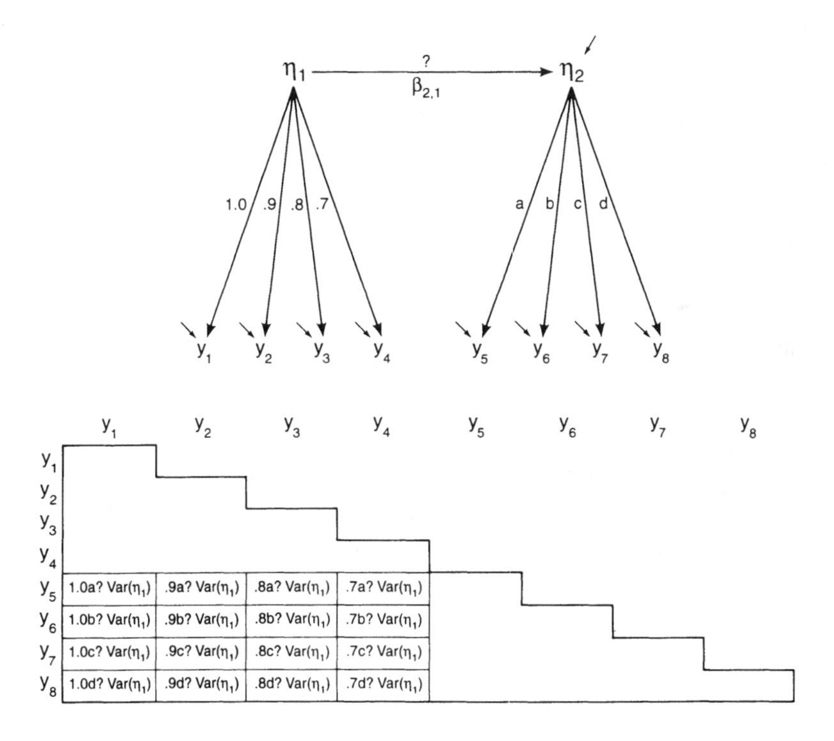

E: Double proportionality from two sets of indicators.

1.3.1 A Pure Factor Model

The third model column of the preface figure details how a common cause of two variables implies a covariance, and hence correlation, between those variables even if the variables have no direct causal effect on one another. This is often called a "spurious relationship." The equation near the bottom of that column informs us that the covariance between the variables sharing the common cause must equal the product of three quantities: the magnitudes of the effects leading from the common cause to each of the two dependent variables, and the variance of the common cause. If we call the dependent variables Ind_1 and Ind_2 and the common cause η, this equation may be written as

$$\text{Cov}(\text{Ind}_1,\text{Ind}_2) = (\text{effect of } \eta \text{ on Ind}_1)(\text{effect of } \eta \text{ on Ind}_2)\text{Var}(\eta). \quad 1.1$$

We can now describe the factor model in Figure 1.2A as consisting of multiple applications of this basic spurious or common cause model. Any pair of indicators selected from y_1 to y_5 can be viewed as the two indicators originating in the common cause, and hence the magnitude of the spurious covariance between each pair of indicators can be calculated by using Eq. 1.1. For example, assuming the diagrammed numerical values represent the magnitudes of the effects of the common cause on the indicators, the covariance between y_1 and y_2 is

$$\text{Cov}(y_1,y_2) = 1.0(.9)\text{Var}(\eta_1). \quad 1.2$$

Similarly, we can calculate

$$\text{Cov}(y_1,y_3) = 1.0(.8)\text{Var}(\eta_1) \quad 1.3$$

and

$$\text{Cov}(y_2,y_4) = .9(.7)\text{Var}(\eta_1) \quad 1.4$$

and so on, for the covariance between any pair of indicators.

We encounter the fundamental proportionality that is of interest when we calculate the covariances of any one indicator with the rest of the indicators. For example, the covariances of the fifth indicator with each of the preceding four indicators are

$$\text{Cov}(y_1,y_5) = 1.0(?)\text{Var}(\eta_1) \quad 1.5$$

$$\text{Cov}(y_2,y_5) = .9(?)\text{Var}(\eta_1) \quad 1.6$$

$$\text{Cov}(y_3,y_5) = .8(?)\text{Var}(\eta_1) \quad 1.7$$

$$\text{Cov}(y_4,y_5) = .7(?)\text{Var}(\eta_1). \quad 1.8$$

Observing that the first numbers on the right-hand sides of these equations are the effects of the common cause η_1 on the corresponding indicators (as depicted in Figure 1.2A), we see that the observed covariances must be strictly proportional to the magnitudes of the effects of the common cause on each indicator. That is, if we use *C* to denote whatever Constant is found to be the product of the values ultimately assigned to the Var(η_1) and the ? effect of η_1 on indicator y_5, we can rewrite these equations as

$$\text{Cov}(y_1,y_5) = 1.0C \quad 1.9$$

$$\text{Cov}(y_2,y_5) = .9C \quad 1.10$$

$$\text{Cov}(y_3,y_5) = .8C \quad 1.11$$

$$\text{Cov}(y_4,y_5) = .7C. \quad 1.12$$

This makes it clear that the covariances of y_5 with η_1's other indicators are proportional to the effects η_1 has on those other indicators.

Here we have an example of a solid, rigid, and uncompromising constraint arising despite a model's inclusion of fictional and flexible components. The "causal effect" of η_1 on y_5 is no less imaginary than any other causal effect, and the magnitude of this effect, denoted ?, is completely free. It may be positive or negative, large or small. Changing the value of ? or the Var(η_1) would change the value of the proportionality constant *C* but for whatever single value *C* ultimately takes on, we *must* observe that y_5's covariances with the *other* four indicators are proportional to the λ effects leading to those indicators. No amount of flexibility in ?'s value, or looseness in the philosophical nature of causality,[6] or in the meaning of η_1 softens the uncompromising, strict, and unavoidable proportionality between y_5's covariances with the other indicators and the effects leading to those indicators.

You may be wondering whether this demonstration has been rigged by prespecification of the 1.0, .9, .8, and .7 effects linking η_1 to the first four indicators. These specific values have merely determined which particular proportionality appears, not whether a proportionality appears. Some particular proportionality will be demanded no matter what λ values are used or estimated. Any effects (small negative, or large positive) could be inserted as η_1's effects on the first four indicators and the strict and uncompromising implication of model 1A would remain that y_5's covariances must be proportional to these effects, with the proportionality constant determined by η_1's variance and η_1's effect on y_5.

1.3.2 A Second Look at the Factor Model in Figure 1.2A

A different perspective on the proportionality of y_5's covariances with the first four indicators is gained if we imagine ourselves considering whether we should add y_5 to a set of four indicators that had been previously factor analyzed. A factor analysis using four indicators and one concept would be small, but estimates for each of the λ loadings and the variance of η_1 could be obtained. We will imagine that the λ values indicated in Figure 1.2A have been thusly acquired, and we now consider adding y_5 into the factor model.

Four new covariances will be added to the data set (the four dotted lines in the lower part of the figure, or the four covariances in the data matrix on the right of the figure) but only one new structural coefficient (the ? value of $\lambda_{5,1}$) is added to the model to assist in modeling these covariances.[7] If we are to maintain the meaning of η_1 determined by the four-variable factor analysis, we cannot change the numerical value of any of the first four variables' loadings or the variance of η_1,[8] and hence the value of $\lambda_{5,1}$ is the only quantity that can be adjusted (by estimation) to account for the four new data covariances. Is it possible for this one model coefficient to account for four data points (covariances)? This is possible, and seeing why it is possible leads us back to the central issue of proportionality constraints.

Consider the constant C which we created as the product of ? and $\mathrm{Var}(\eta_1)$. Equations 1.5-1.8 (or 1.9-1.12) inform us that if the five-indicator model is true, the covariances between y_5 and the previous four indicators must equal 1.0, .9, .8 and .7 times this value. We could use the first of either of these sets of four equations to calculate the required value of C. Specifically, the covariance between y_5 and y_1 is available from the data set, and we previously estimated the loading of y_1 on η_1 as 1.0, so $C = \mathrm{Cov}(y_1,y_5)/1.0$, or simply $C = \mathrm{Cov}(y_1,y_5)$. Using this value of C in the subsequent three equations provides the values of the remaining three y_5 covariances that must be observed if the five-indicator factor model is to be consistent with the data.

$$\mathrm{Cov}(y_2,y_5) = .9\mathrm{Cov}(y_1,y_5) \qquad 1.13$$

$$\mathrm{Cov}(y_3,y_5) = .8\mathrm{Cov}(y_1,y_5) \qquad 1.14$$

$$\mathrm{Cov}(y_4,y_5) = .7\mathrm{Cov}(y_1,y_5). \qquad 1.15$$

That is, y_5's covariances with y_2, y_3, and y_4 must be specific fractions (or multiples[9]) of the covariance between y_5 and y_1, and the specific fractions are[10] the loadings obtained from the four-indicator factor analysis. Hence, the proportionality between the λ's and y_5's covariances discussed in the preceding section appears as a demand for specific relative sizes among y_5's covariances.

If the behavior of the five y variables in the real world is indeed controlled by the operation of a common causal force, this strict patterning of relative covariance magnitudes will be observed in the data, and the feat of fitting four covariances with a single coefficient will have been accomplished. (The required value of ? could be calculated by rearranging any one of Eqs. 1.5 to 1.8 since $\mathrm{Var}(\eta_1)$ would be known from the previous four-indicator factor analysis.)

But the forces in the real world may function differently. For example, one of the y's may influence y_5 in addition to displaying a common dependence on η_1, so that the covariance between y_5 and this y is more than spurious, or some additional factor may be common to y_5 and only some of the first four y's. In such cases the strict proportionality between the λ loadings and the y_5's covariances will not be observed, and the model will fail to match the data.[11] Note again that the uncompromising demand for proportionality between the λ loadings and y_5's covariances (here expressed as a demand for some particular proportionality between the observed y_5 covariances) cannot be circumvented or avoided by providing even unlimited flexibility in the value assigned to η_1's effect on y_5, or altering the meaning of the concept η_1.

1.3.3 A Nearly Factor Model

We are now prepared to move beyond the factor model to consider Figure 1.2B. Here we will also ultimately re-encounter the proportionality phenomenon, but we cannot start from Eq. 1.1 because we have two concepts (factors), not one as required by this equation. We could get the model-implied covariance between y_5 and each of the previous four indicators by specifying the full 1.2B model in matrix form and using *Essentials* Eq. 4.50 or 4.62, but this would have the disadvantages of demanding full and formal specification of the model (which would waste space) and hand calculation of an inverse (which can be awkward). It is easier to examine the particular covariances of interest by using a procedure that accomplishes the same thing as the general formulas but that focuses on one covariance or variance at a time, rather than on the full covariance matrix.

The required procedure was used in *Essentials* to derive the equations we adopted as Eq. 1.1 above, and one finds it employed repeatedly in the literature, so it is well worth learning. The procedure is discussed and illustrated in Duncan (1975:13, 29), so we merely summarize the necessary steps.[12]

If one is interested in a model's implications for a particular covariance:

1) obtain the model's equation for each of the two variables involved in the covariance;

2) use these equations to write a new equation having the product of the two variables on the left side of the equal sign and the product of the multiple component parts equivalent to each of the variables on the right side;

3) take expectations of both sides of the equation;[13]

4) rewrite the left side as the desired covariance (since this is the definition of the appropriate covariance, assuming the means of the variables are zero, *Essentials* Eq. 1.8) and simplify the right side of the equation by first multiplying out the terms, then moving the expectation operator (averaging) within the summed parts, and simplifying wherever possible.[14]

Let us apply this procedure to see what the model in Figure 1.2B implies about the covariance between y_1 and y_5. The equations for y_1 and y_5 are

$$y_1 = 1.0\eta_1 + \epsilon_1 \qquad 1.16$$

$$y_5 = 1.0\eta_2 + \epsilon_5. \qquad 1.17$$

The product equation is

$$y_1 y_5 = (1.0\eta_1 + \epsilon_1)(1.0\eta_2 + \epsilon_5)$$

$$= 1.0(1.0)\eta_1\eta_2 + \epsilon_1\eta_2 + \epsilon_5\eta_1 + \epsilon_1\epsilon_5. \qquad 1.18$$

Taking expectations provides

$$E(y_1 y_5) = E(1.0(1.0)\eta_1\eta_2 + \epsilon_1\eta_2 + \epsilon_5\eta_1 + \epsilon_1\epsilon_5) \qquad 1.19$$

$$= 1.0(1.0)E(\eta_1\eta_2) + E(\epsilon_1\eta_2) + E(\epsilon_5\eta_1) + E(\epsilon_1\epsilon_5). \qquad 1.20$$

The left side of this equation is the definition of the covariance between the variables y_1 and y_5 (assuming the variables are recorded as deviations from their means so the $E(y_1) = E(y_5) = 0.0$, *Essentials* Eqs. 1.6, 1.8). The assumed independence of the error variables from the predictor variables in the equations, and from one another, makes the three right-most terms zero (*Essentials*, 20). We are left with

$$Cov(y_1 y_5) = 1.0\ (1.0)E(\eta_1\eta_2) + 0 + 0 + 0$$

$$= 1.0(1.0)Cov(\eta_1\eta_2) \qquad 1.21$$

$$= 1.0(?)1.0$$

in which a ? has been inserted to highlight the unknown value of $Cov(\eta_1\eta_2)$.

Repeating these steps for the covariances between y_5 and the other indicators gives

$$\text{Cov}(y_2y_5) = .9(1.0)\text{Cov}(\eta_1\eta_2) = .9(?)1.0 \qquad 1.22$$

$$\text{Cov}(y_3y_5) = .8(1.0)\text{Cov}(\eta_1\eta_2) = .8(?)1.0 \qquad 1.23$$

$$\text{Cov}(y_4y_5) = .7(1.0)\text{Cov}(\eta_1\eta_2) = .7(?)1.0. \qquad 1.24$$

You should immediately see that these equations parallel Eqs. 1.5-1.8, and 1.9-1.12, if we view the covariance between η_1 and η_2 as being the ? whose unknown value is to be estimated, and the constant 1.0 as taking the place of the constant Var(η_1). This convinces us that model 1.2B also demands a strict, rigid, and uncompromising proportionality between the λ's linking η_1 to its indicators and y_5's covariances with those indicators. If the data do not cooperate by following this proportionality, the model would fail, just as the factor model in Figure 1.2A would fail.

1.3.4 Nonfactor Models

It is now but a short step to Figures 1.2C and 1.2D. These figures are similar to Figure 1.2B, except that they specify the source of the covariance between the concepts η_1 and η_2 as being the result of a direct effect of one of these on the other. Let us begin by using Duncan's four-step procedure to obtain model 1.2C's explanation of (or implication structure providing for) the covariance between η_1 and η_2. We write the equations,

$$\eta_1 = \eta_1 \qquad 1.25$$

$$\eta_2 = \beta_{2,1}\eta_1 + \zeta_2 \qquad 1.26$$

multiply, take expectations,

$$E(\eta_1\eta_2) = E(\beta_{2,1}\eta_1\eta_1 + \zeta_2\eta_1) \qquad 1.27$$

and simplify, by assuming the error on η_2 is independent of η_1,

$$\text{Cov}(\eta_1,\eta_2) = \beta_{2,1}\text{Var}(\eta_1). \qquad 1.28$$

If we were mathematical purists, we would now repeat for model 1.2C the steps that gave us model 1.2B Eqs. 1.21-1.24, but since the steps and results are identical, we will leave them to you, and merely use this fact by substituting the right side of Eq. 1.28 in place of the Cov($\eta_1\eta_2$) in Eqs. 1.21 through 1.24.

$$\mathrm{Cov}(y_1,y_5) = 1.0(\beta_{2,1})\mathrm{Var}(\eta_1) = 1.0(?)\mathrm{Var}(\eta_1) \quad 1.29$$

$$\mathrm{Cov}(y_2,y_5) = .9(\beta_{2,1})\mathrm{Var}(\eta_1) = .9(?)\mathrm{Var}(\eta_1) \quad 1.30$$

$$\mathrm{Cov}(y_3,y_5) = .8(\beta_{2,1})\mathrm{Var}(\eta_1) = .8(?)\mathrm{Var}(\eta_1) \quad 1.31$$

$$\mathrm{Cov}(y_4,y_5) = .7(\beta_{2,1})\mathrm{Var}(\eta_1) = .7(?)\mathrm{Var}(\eta_1). \quad 1.32$$

The numerical value of the variance of η_1 will be determined by η_1's four indicators, and the effect $\beta_{2,1}$ will function as our new ?, so we again find proportionality demands similar to those of models 1.2A and 1.2B.

To shorten the story, we leave to you the demonstration that in model 1.2D, Duncan's procedure says our model accounts for the covariance between η_1 and η_2 by,

$$\mathrm{Cov}(\eta_1,\eta_2) = \beta_{1,2}\mathrm{Var}(\eta_2) \quad 1.33$$

and Duncan's procedure applied to the covariances between y_5 and the other indicators again provides exactly the Eqs. 1.21-1.24. Inserting the right side of the $\mathrm{Cov}(\eta_1,\eta_2)$ equation into these equations again gives a set of four equations expressing a strict proportionality demand between y_5's covariances and the λ's leading from η_1 to its indicators. The $\beta_{1,2}$ now functions as the ?, and η_2's variance must be well determined by "fixing" it or the corresponding measurement error variance, as discussed in *Essentials* (118-122).

$$\mathrm{Cov}(y_1,y_5) = 1.0(\beta_{1,2})\mathrm{Var}(\eta_2) = 1.0(?)\mathrm{Var}(\eta_2) \quad 1.34$$

$$\mathrm{Cov}(y_2,y_5) = .9(\beta_{1,2})\mathrm{Var}(\eta_2) = .9(?)\mathrm{Var}(\eta_2) \quad 1.35$$

$$\mathrm{Cov}(y_3,y_5) = .8(\beta_{1,2})\mathrm{Var}(\eta_2) = .8(?)\mathrm{Var}(\eta_2) \quad 1.36$$

$$\mathrm{Cov}(y_4,y_5) = .7(\beta_{1,2})\mathrm{Var}(\eta_2) = .7(?)\mathrm{Var}(\eta_2). \quad 1.37$$

We will not discuss Figure 1.2E in detail, but merely draw your attention to a double proportionality implied by this model. Each of the rows in the highlighted portion of the covariance matrix could be treated as if it arose as the y_5 in our discussion of model 1.2C. Hence each row in the highlighted portion of the matrix will display a strict proportionality to the λ's linking η_1 to its indicators.

The columns in the highlighted area must also display a proportionality. This can be demonstrated by working through the mathematics parallel to model 1.2D (with the identity of the concepts reversed). In Figure 1.2E we have labeled η_2's λ's as *a*, *b*, *c*, and *d* to highlight that each of these λ's falld in a particular row and so provide a proportionality between rows. Hence, with multiple indicators for two concepts, the model implies we must observe one proportionality within the rows and a second proportionality within the columns, in the "between concept" portion of the observed indicator covariance matrix. This is a strong demand, and its nonattainment has led to the failure of many a multiple-indicator model.

What are we to make of these mathematical ramblings? First, *proportionality constraints are not unique to factor models, they pervade any model that postulates multiple indicators.*[15] The factor model is neither uniquely endowed nor deficient in this respect.

Second, models with multiple indicators are prone to failing because of the stringency of their proportionality demands. *The more variables with multiple indicators, the greater the number of proportionality demands*, and hence the more potentially embarrassing challenges the data can provide.

Third, *there is no mathematical preference for proportionalities arising from indicators grouped under a single concept to proportionalities arising between concepts*, or vice versa. Both styles of proportionality requirements can provide the sting of a model's failure, or the glow of predictive success. This has many implications. For starters, we must acknowledge that evidence on how a concept links to another concept can be as devastating to the claim that several indicators share a common source as is adding another indicator of that source. Model 1.2A adds another indicator of the postulated common source η_1, while models 1.2B, C, and D link that same supposed source to an indicator of a conceptually distinct factor η_2. *The simultaneous demands for a dependence of indicators y_1 to y_4 on the original concept η_1, and for a connection between the concepts η_1 and η_2, place proportionality constraints on the covariances between the indicators of those concepts, and hence test η_1's ability to function as a unitary source coordinating the behavior of indicators y_1 to y_4.*

Anderson and Gerbing (1988:415) put it this way: "Because it often occurs in practice that there are less than four indicators of a construct, external consistency then becomes the sole criterion for assessing unidimensionality." The terms "external consistency" and "external validity" are bits of jargon encapsulating the demand for the proportionality of covariances when external causes or effects are linked to the variable the researcher has been myopically focusing on as "the" concept to be measured. A set of indicators are unidimensional if they indeed arise from a single common source as in Figure 1.2A. We return to the issue of unidimensionality in Section 2.2.3.

Factor analyses having a single factor (like 1.2A) offer no choice but to use proportionality demands arising from internal consistency to evaluate the adequacy of the connections between the concept and its indicators. Factor analyses allowing two or more correlated factors (like 1.2B if η_2 had more indicators, or 1.2E if the link between the η's was a correlation instead of an effect) incorporate double-proportionality demands in their model assessment but such models are plagued by a couple of entrenched operating strategies that are based on outdated advice.

Factor analysis has historically been justified as a pursuit of parsimony, and hence the standard advice is to minimize the number of factors. But it is illusory to believe that the fewer the concepts, the more parsimonious is one's theoretical world. As one adds more and more concepts (with a fixed number of indicators), the equally standard policy of estimating a full matrix of free correlations among the factors becomes a direct assault on parsimony. As the number of factors increases, the number of covariances among those factors increases even more rapidly, and parsimony is the innocent victim. Hence the standard admonition to avoid angering the statistical gods (risking an underidentified model) by seeking a minimum number of factors.

From the broader perspective of structural equation modeling, *this amounts to a bias against ever seeing that equally parsimonious models can be achieved by imbedding more concepts (factors) in sparser causal networks*. I see no reason to believe there is a divine preference for models with few fictional concepts and many fictional correlations among them, over models having more fictional concepts connected by fewer fictional (effect) relations.

The second standard operating policy, that of selecting items from a limited domain of content, is even more disturbing. Consider Figure 1.2B again. Is there anything that keeps me, or you, from identifying y_5 as age, attitude, income, weight, or any other easily measured indicator? In fact, any other correlated variable will do. But have you ever seen a published example of someone including an obviously "out of domain" indicator like age in a factor analysis? The advice to include only items sharing common sources biases item selection specifically to avoid merely correlated variables (as in 1.2B) or causal pre- or postcursors (models 1.2C and D). There are a plethora of mundane variables routinely available; why are they not entered into factor analyses?

There are many reasons for this advice; all well-meaning but most unhelpful. One standard piece of advice to those doing factor analysis is that you must have three or four indicators of a concept before it should be entered into your factor analysis. Suppose I now take the participant's age as the mundane variable being considered for inclusion because I wish to use its covariance constraints to test the subsumption of the first four indicators under the concept η_1. Where can I get four indicators of your age? Ask you four times? Ask you and three of your friends?

What is it that prompts the advice to use three or four indicators? Those proffering this advice are actually trying to be helpful. Following this advice almost guarantees that the selected model is identified and hence estimable, even with all the factor covariances free. The problem is, this advice does not take into account that one may not have to estimate as many coefficients if one is thinking in terms of substantive models incorporating effects among the concepts, instead of permitting all the possible correlations. Why invite problems by estimating all the possible correlations among the concepts (factors) if one has a sparse structural model in mind?[16]

We should not be too hard on factor analysts about this. After all, they are working in a tradition where the concepts are supposedly very abstract (as if there are degrees of fictionality). Their mistake is merely in thinking that all the concepts we work with must be equally recalcitrant. They simply have not considered the progressive models depicted in Figure 1.2, and the advantages of purposefully selecting y_5 to be an indicator that is as clear, as nondebatable, and as easily and directly measurable as possible.

But neither should we be exceptionally tolerant of those refusing to modify their standard operating practices. Researchers who steadfastly refuse to include measures of suspected causes or effects of the very abstract factor they are trying to measure should be called to task for purposefully avoiding a relatively easy test of their model. We have just seen that any cause or effect variable places stringent proportionality constraints on the data, and hence allows for the data to decisively contradict a claim of "common cause." Anyone purposefully avoiding such easy and obvious tests should be ridiculed for their purposeful evasiveness. In disciplines like psychology, where factor analysis is entrenched, it may take a generation of young upstarts to repeatedly point this out before we see a marked modification of established traditions.

In summary, there is considerable similarity in how both factor models and multiple indicator structural models account for the observed covariances/correlations among items. These modeling styles share a common mathematical base, and they share a theoretical commitment to both latent factors/concepts and to causation. The key difference is that the factor model's primary[17] explanatory mode is "spurious item correlations (covariances) arising from a dependence on the common-factor cause," while structural equation modeling permits explanations based on postulation of causal effects among the concepts, and may or may not incorporate multiple indicators. Structural models incorporating multiple indicators are similar to factor models in that they imply a strict proportionality between the indicator loadings (λ's) and the indicator covariances with each and every other modeled variable. There is no mathematical reason to inherently prefer one style of model over the other. Both are acceptable, and in fact are mathematical cousins.

In Chapter 2 we encounter a suggestion that one of these styles (the factor model) should be treated as more fundamental or basic in that it should be routinely investigated first. The demonstration of the strong parallels between multiple indicator models and factor models prepares us for this debate by sharpening the distinction between acceptable and unacceptable justifications for any such preference. Any preferential treatment cannot be based on the contribution of proportionality constraints to overall model tests, or to the identification of estimates, because both modeling styles incorporate similar proportionality demands. Hence, the justification for any preference must be found elsewhere, namely among any remaining differences between these modeling styles. Substantial differences remain, and we will spend considerable time examining these differences as we discuss the two-step approach. Before entering this particular fray, however, we need to be clear about one further fundamental claim implicit in factor analysis that has misled more than a few researchers, namely the necessarily proximal nature of factors.

1.4 The Necessarily Proximal Nature of Factors

Let us begin with an assertion about Figure 1.3A. *η_1 is a common cause of the y indicators but it is not a common factor for those indicators.* Only heresy could lead one to question whether a common cause might not be a common factor. Only simple mathematics will delay my excommunication.

The first half of the mathematical argument comes from focusing on the covariance between y_1 and y_2 in Figure 1.3A. From Eq. 1.1 we know that this covariance must be

$$\text{Cov}(y_1,y_2) = \lambda_{1,2}\lambda_{2,2}\text{Var}(\eta_2). \qquad 1.38$$

Now consider η_2 as a dependent variable caused by η_1 and an independent error variable ζ_2. We note that the equation for η_2, namely $\eta_2 = \beta_{2,1}\eta_1 + \zeta_2$ is of the same general form as the equation heading column two of the preface figure, and hence the equation in the lower segment of this column informs us about the sources of η_2's variance. Specifically, since the error variable is independent of the causal variable η_1, we can partition η_2's variance into explained and error variance as follows:

$$\text{Var}(\eta_2) = \beta_{2,1}^2\text{Var}(\eta_1) + \text{Var}(\zeta_2). \qquad 1.39$$

Substituting this equation for the Var(η_2) in the preceding equation, we find that model 1.3A implies that the covariance between indicators y_1 and y_2 equals

$$\text{Cov}(y_1,y_2) = \lambda_{1,2}\lambda_{2,2}(\beta_{2,1}^2\text{Var}(\eta_1) + \text{Var}(\zeta_2)) \qquad 1.40$$

which we will rewrite, for later comparison purposes, as

$$\text{Cov}(y_1,y_2) = (\lambda_{1,2}\beta_{2,1})(\lambda_{2,2}\beta_{2,1})\text{Var}(\eta_1) + \lambda_{1,2}\lambda_{2,2}\text{Var}(\zeta_2). \qquad 1.41$$

For the second part of the mathematics, we turn to the factor model in Figure 1.3B which depicts a variable η_1^* that is like η_1 in that the effects it has on the y indicators are precisely the indirect effects η_1 has on those indicators in model 1.3A. That is, each effect here is the product of the direct effects comprising the corresponding indirect effect in 1.3A. Hence, using Eq. 1.1

Figure 1.3 Only Proximal Factors.

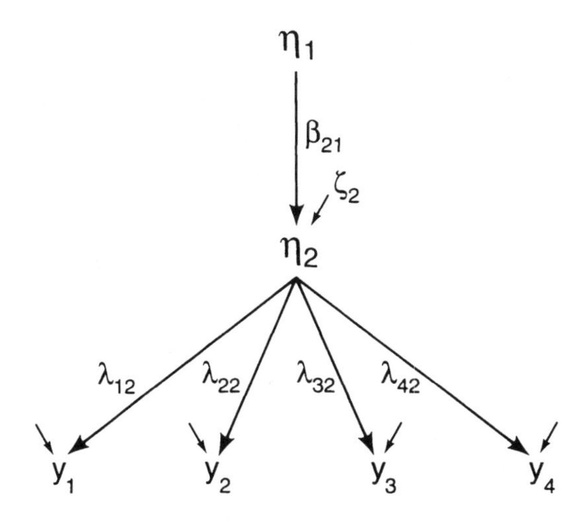

A: The real world.

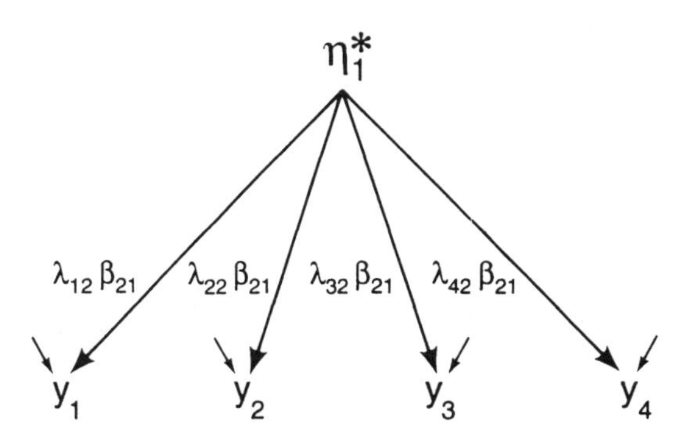

B: Trying to use η_1 as a factor.

again, we can write the covariance between y_1 and y_2 in model 1.3B as

$$\mathrm{Cov}(y_1,y_2) = (\lambda_{1,2}\beta_{2,1})(\lambda_{2,2}\beta_{2,1})\mathrm{Var}(\eta_1^*). \qquad 1.42$$

The covariance between y_1 and y_2 is fixed by the data, so the left sides of Eqs. 1.38, 1.40, 1.41, and 1.42 are identical. The right side of Eq. 1.38 tells us that *all* of the variations in η_2 contribute to the covariance between the indicators, through the effects η_2 has on the indicators.[18] Equation 1.40 clarifies this by specifically acknowledging that the variations in η_2 arising from both η_1 *and the error variable* ζ_2 contribute to the indicator covariance. Equation 1.42 describes model 1.3B and highlights the problem.

The variable η_1^* is like η_1 (in Figure 1.3A) in that we forced η_1^* to display the same causal effectiveness that η_1 has on the y indicators. But η_1^* *could not possibly be* η_1 *because it must have a variance that differs from the variance of* η_1. If the variance of η_1^* equalled the variance of η_1, the right side of Eq. 1.42 would be identical to the first term on the right of Eq. 1.41, but this is impossible because this would claim that $\mathrm{Cov}(y_1,y_2)$ must equal two quantities that are themselves unequal. $\mathrm{Cov}(y_1,y_2)$ would have to equal the right side of 1.42 or the right side of 1.42 plus the non-zero value $\lambda_{1,2}\lambda_{2,2}\mathrm{Var}(\zeta_2)$ as seen in 1.41. This is impossible, so we must conclude the variance of η_1^* cannot equal the variance of η_1, even though η_1^* has the same effects as η_1.

This means the variables η_1 and η_1^* cannot be the same variables! A variable has only one variance, so we must conclude η_1 and η_1^* are different variables even though we tried to force them to be the same by forcing them to display identical effects! The variations in η_2 arising from the error source contribute to the covariance between the indicators, but this part of η_2's variance is ignored if the η_1^* in model 1.3B is the same as the η_1 in model 1.3A.

This informs us that even though η_1 is a common cause of all of the indicators, it is not the factor that would be found if one did a usual factor analysis, or even a confirmatory factor analysis that had assisted in identifying the factor by using fixed loadings equivalent to the proper indirect effects (as in our Figure 1.3B). In fact, estimating a factor model with free loadings would actually locate a variable identical to η_2. The estimated loadings would equal the λ values in model 1.3A and the variance estimated for the concept would equal $\mathrm{Var}(\eta_2)$.

These observations carry the strong conclusion that *even though a specific variable is named and justifiably identified as a common cause of a set of indicators, this is insufficient to conclude that "this" is the variable that will be located as the common factor in a factor analysis*. Factor analyses will locate only the common cause precisely at the branching that sends unique causal impacts to each of the indicators.[19] *This means we must routinely request some additional justification or evidence regarding the identity of factors beyond the correct labeling of a factor by a title identifying it as a common cause of the items. One must be able to provide some evidence as to why the identity (name) provided for the variable should correspond to the most proximal of the potentially many*[20] *common causes of the items.*

This will seem disconcerting to some factor analysts. We are asking for a style of evidence antithetical to common cause thinking. The practitioner is being asked to defend the assertion that each of the effects leaving the common factor and heading towards the indicators necessarily functions by separate, non-overlapping, unique, unmixed, unshared, different, and distinct causal mechanisms. These are words factor analysts are accustomed to hearing as descriptions of the error variables on the indicators (uniquenesses), but we are asserting that this style of description is a necessary component of the description of the λ values or loadings linked to the common factor. To guarantee that the researcher is mentally pointing to η_2 and not η_1 (in Figure 1.3A) as the common factor, we must request that the researcher explicitly acknowledge, justify, and account for the uniqueness of the causal forces emerging from the common factor and taking separate routes to the indicators. No common causal segments (like $\beta_{2,1}$) are permitted. This is no minor task for researchers accustomed to feeling good about pointing to the commonness of the common factor as the foundation of their identification of that factor. Note also, that this demands a consideration of directed effects among the variables potentially designatable as "the factor," and therefore also runs afoul of factor analysts' penchant for thinking of factors as merely correlated entities.

1.4.1 Devious Distal Factors

It is easy to be led astray by inattention to factor proximality as a concern in modeling common causes. I know, I have been fooled by this. What is worse, and somewhat embarrassing, is that if you have read *Essentials*, I have probably also misled you because of this!

The mistake I made because of inattention to the proximal nature of factors did not occur in the context of factor analysis, but in the discussion of acceptable model simplifications (*Essentials*, 150-154). At that point I was seeking acceptable model simplifications, in the hopes of finding ways to improve the identification status of models. I ultimately concluded that model simplification is of limited use in improving identification, and correcting the mistake leads to an even more pessimistic conclusion, so I need not repeat the whole discussion here. For our current purposes, it will be sufficient to develop a sense that the mandatory proximal nature of factors is a problem which transcends the bounds of traditional factor analysis.

Figure 5.7 in *Essentials* depicts several basic models, and a series of proposed simplifications of those models, along with an assertion of whether

Figure 1.4 Correcting *Essentials* Figure 5.7.

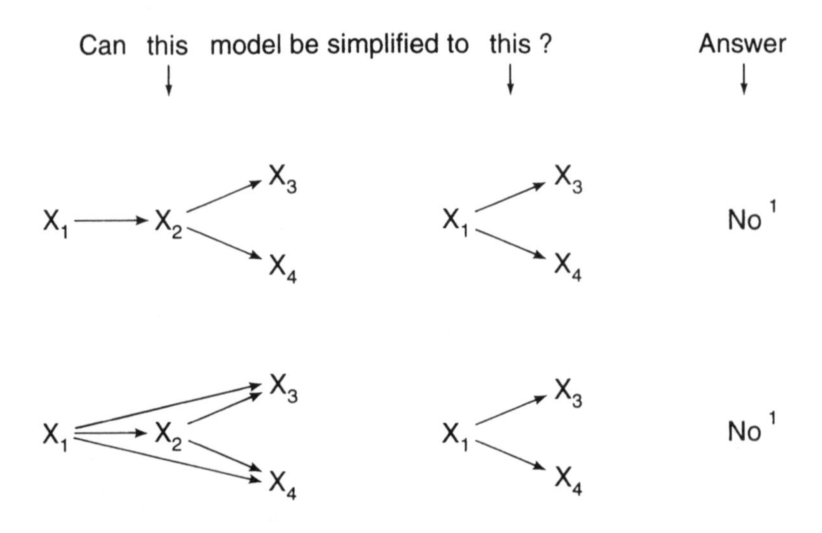

Note: The following footnote is to appear in the sixth printing of *Essentials*. Though the covariance between X_3 and X_4 continues to be modeled as spurious, this simplification is unacceptable because it mistakenly claims that X_1 is the entire source of the spurious covariance. If X_2 is less than perfectly accounted for by X_1 then there is error variance in X_2, and this error variance contributes a component of the covariance between X_3 and X_4. *Essentials* Eq. 1.67 implies that the covariance between X_3 and X_4 equals the product of the effects of X_2 on X_3 and X_4, and the variance of X_2. But the variance of X_2 is partially due to X_1 and partially due to other sources (error), so the covariance is correspondingly partitioned and hence should not be depicted as solely due to X_1, as it is in this simplification. This was incorrectly reported as a Yes in the first five printings of *Essentials*.

each simplification is acceptable or not. Two of the instances where I report "Yes" the simplification is acceptable, should report "No," indicating the simplification is unacceptable. The two offending cases are represented in Figure 1.4. The problem is that there is an implicit error variable influencing X_2 in both of these models, and hence the error variance in X_2 contributes to the covariance between X_3 and X_4 in a direct parallel to the discussion of Eqs. 1.40 and 1.41 above. By failing to acknowledge this contribution, the simplified models will attribute all the covariance between these variables to X_1 and hence will produce biased estimates of the total effects of X_1 on X_3 and X_4. The estimated effect of X_1 omitting X_2 could not equal the indirect effects of X_1 through X_2 in the basic model.

At least I can say that I found this problem myself, and that I intend to correct this in the next printing of *Essentials*.[21]

1.5 How to Model Multiple Indicators

Given all the above, how should one proceed when modeling multiple indicators in LISREL models? Let me first summarize and then elaborate upon the answer implicit in *Essentials*.[22] My advice was to begin by narrowing down the number of potential indicators by focusing on the best two, or three, indicators of your concept. These assessments were to be based on the methodology that provided the data (i.e., clear and appropriate wordings of questions, appropriately scaled answer categories, sufficient variance, etc.). Then:

1) choose the indicator you believe is the single best available indicator of the concept in question,
2) fix the λ for that indicator at 1.0,
3) fix the θ variance for that indicator at a specific value (see below), and
4) enter free λ and θ variances for the second and third best indicator(s) of the concept, unless you wish to be extra assertive by entering fixed or constrained values here also.

The first three steps are designed to provide the underlying concept a clear and unchanging meaning. Step one informs your readers about which part of the shared external world (the indicators) is most similar to your fictional η. Step two sets the scale for that concept to correspond to the scale for the observed indicator, by making each real unit of change in the concept correspond to exactly one unit of change in the indicator.

Step three quantifies your assessment of how similar or dissimilar your concept is to the best indicator. This assessment is under your control because you are free to change the meaning of your conceptual fiction (your η), and hence it is you who controls the size of the gap between your concept and any given indicator. Once you have located a single meaning for your concept, you can most clearly inform others of that meaning by specifying precisely how close that meaning is to the best of the indicators by specifying (fixing) the θ variance for that indicator.[23] Altering the value of the θ variance changes the meaning of the concept because it claims there is more or less similarity between this indicator and your concept. Given that the indicator is not changing, any widening or narrowing of the concept-indicator gap implies that it must be the meaning of the concept that is changing, and this is not permissible if the model is to reflect a single theory.[24]

The fourth step is possible only if there are two or more indicators of the concept, and it provides for a test of the conceptualization you have asserted via steps one through three. That is, without multiple indicators, the best you can do is to clearly inform your readers about your conceptualization. With two or more indicators you can both inform your readers about your conceptualization and test that conceptualization, via your model's requirement of proportionality constraints, as discussed above.[25]

Figure 1.5 depicts the essence of this strategy. Here y_1 is the indicator most closely resembling the implied conceptualization of η_1 and the fixed 1.0 sets the scale of η_1 by guaranteeing that each unit change in η_1 results in a corresponding unit change in y_1. Note that not all changes in y_1 necessarily result from changes in η_1. If there were zero error variance for y_1 the only changes in y_1 would originate in η_1 but y_1's error variance typically is not zero because other variables may influence y_1. This implies that while the η_1 and y_1 scales correspond, η_1's values display less variance or are less spread out along that scale, than are y_1's values.[26]

If I know what my η_1 concept means, that is, if my conceptualization is fixed and does not aimlessly wander to meanings closer to or further from y_1, I can most clearly tell you about my concept's meaning by honestly assessing and reporting the degree of correspondence or discordance between η_1 and y_1 (via fixing the error variance $\theta_{\epsilon 1}$). It is my image of η_1 and *not* the real world (y_1) which determines the proximity between η_1 and y_1. Fixing the error variance assumes you have had sufficient experience with the methodology you are employing to make some reasonable assessment of the likely sources of interference affecting the best indicator. It also assumes you have a commitment to a specific conceptual meaning for η_1. If you initially do not have a specific meaning for η_1, this procedure forces you to develop one, and hence forcibly sharpens your conceptual precision.

Contrast this with the routine freeing of $\theta_{\epsilon 1}$, where a new meaning is assigned to a similarly labeled concept each time a new data matrix is used and a new $\theta_{\epsilon 1}$ is estimated. At this point I cry: flaccid theory! Freeing $\theta_{\epsilon 1}$ loosens the meaning of η_1 and reveals the researcher's lack of commitment to a particular, even if fictional, meaning for η_1. If η_1 has one meaning there is only one appropriate error variance for y_1. It is only if the meaning of concept η_1 is both fictional *and loose* that y_1's error variance becomes free.

If the estimated proportion of error from a first data set is used to determine a fixed value in subsequent analyses, this problem is reduced but not removed. One has adopted step three of the suggested procedure but with some slightly questionable baggage. One would want a reason for why the meaning of η_1 from a particular data set should become the standard meaning, instead of

Figure 1.5 The Specification of Multiple Indicators.

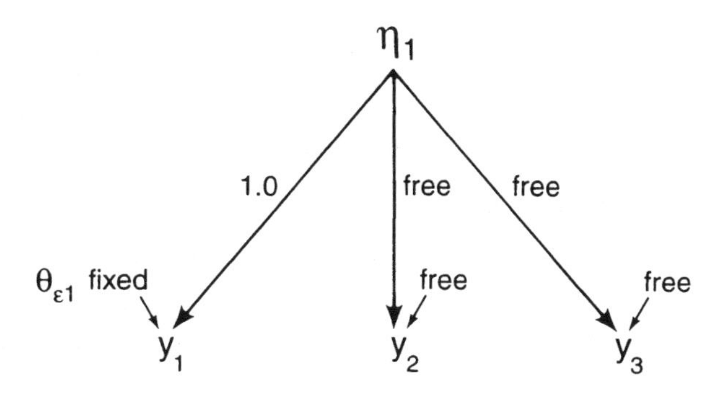

the theorist's standard meaning.

Modeling one indicator of a concept with a 1.0 λ and a fixed θ variance, and a second indicator of that concept with a free λ and θ, commits the researcher to claiming that the entire nonrandom (nonindependent) part of the second indicator arises from its link to the concept whose meaning is already rigidly specified by the first indicator. The residuals for the multiple indicators contribute to the model χ^2 test but the substantive test is whether the λ estimate is large enough, and the θ_ϵ estimate small enough, for the researcher to feel comfortable with this degree of connection between the concept and the second indicator. If the best indicator is assigned 5% error variance, and the λ estimate for a nearly identical second-best indicator results in a 30% error estimate, the researcher should cringe with a pang of theoretical spasm because the meaning for the concept imbedded in the final model is unexpectedly closer to the first than the second indicator. Even an acceptable model χ^2 should be insufficient to hide the resulting theoretical flinch from attentive academic observers.

My research style is to object to loose fictions (free $\theta_{\epsilon 1}$). *I see no way to avoid the fictional nature of abstractions, but I do see a way to avoid imprecision or looseness in our abstractions. Namely, fix the error variance for the scaling indicator at the value corresponding to the researcher's precise fiction.* If one's conceptualization really is loose, and one does not have a single fixed meaning for η_1, then go ahead and leave the error variance for y_1 free, but be aware that multiple indicators become a necessity, and that the indicators provide a weaker test because some part of their covariance information is being used to estimate the meaning of η_1. You should anticipate that the initial loose meaning of your concept will be seen by all, and that this looseness will have been completely eradicated as soon as the error variance is estimated. If the η_1 conceptual chameleon you present in introducing such a loose model is not monochromatically incarnated in your discussion of the estimated model, reviewers will obligingly point out your theoretical color blindness.

Now that your meaning is vividly clear, reviewers are likely to disagree with your specific conceptualization. This is desirable from a disciplinary perspective, and it is also desirable from your perspective as researcher if this sets you up as the key purveyor of a particular conceptualization. But it also has the potential for unwarrantedly shortening one's vitae if the reviewer's disagreement is grounded in trivial conceptual wrangling. Fortunately, one can forestall trivial disagreements by assessing and reporting the sensitivity of your substantive findings to your specific error specification.

The strategy is to provide a footnote summarizing the changes to the substantive parts of the model that arise when you do a series of runs in which each of the fixed θ values is changed to first half and then double the value used in the basic model.[27] This provides a description of how sensitive your substantive conclusions are to the specification of your precise conceptualization. My experience is that if the best indicator has been selected, halving and doubling the measurement error variance is not very disruptive because the bestness of the indicators implies that one is halving or doubling a relatively small quantity. From the referee's perspective, this indicates that minor differences in conceptualization can be tolerated without any substantive change in findings. Consequently, only substantial meaning differences will be aired, and if you end up clarifying your discipline's substantive conceptualizations, even at a reviewer's prodding, it is your vitae that will expand.

By entering a fixed θ for the best indicator, one is in effect adjusting for a specific amount of measurement unreliability. This adjusting is similar to correcting for attenuation (Zeller and Carmines, 1980:63-65) and the more error one inserts, the larger the estimated structural effects are likely to be. That is, it takes a stronger structural effect to display itself in light of larger measurement errors. Lest one be enticed to create strong structural effects by inserting unwarrantedly strong measurement error variances, note that any such gain is doubly paid for. You purchase this at the cost of admitting your methodological ineptitude (you are not in sufficient control of your methodology to obtain any indicator that is similar to your concept) and you are inviting estimation problems.

If the concepts in your theory become too loosely connected to the indicators, the model may become underidentified. And *an underidentified model is an underidentified theory*! Just as an underidentified model indicates the combined constraints of the data and model are insufficient to determine the effectiveness of the variables in the model,[28] an underidentified theory indicates that the combined constraints of the data and theory are insufficient to determine (inform us about) the effectiveness of the variables in the theory. An underidentified theory informs us that any assessments of effectiveness the theoretician has formed on the basis of this theory are untrustworthy because these are essentially a random choice from the many alternative effect estimates capable of reproducing the observables in the theorist's world. Consequently, we should distrust this theorist's claims that specific effects are part of this theory because no specific magnitudes of effects are derivable from this particular theory even with the help of all the covariance data.

Conversely, should especially small θ's be postulated? I recommend not. Even if some statistical benefits accrue from narrowing the concept-indicator gap, this would not lead me to recommend that researchers routinely abandon their current conceptualization of η_1 in favor of something closer to y_1 by minimizing $\theta_{\epsilon 1}$. The researcher's own conceptualization should serve as the starting point. Whether that conceptualization is close to, or far from y_1, we start there because this maximizes the researcher's understanding of the model, enhances the researcher's commitment to the model, and makes the model the most accurate embodiment of this researcher's theoretical position (recall Section 1.1).

The estimated λ and θ values for the additional indicators are metric coefficients and consequently provide scale-dependent statements of the quality of these indicators. The squared multiple correlation corresponding to the θ estimate is usually more informative, with a low R^2 being problematic.[29] The λ estimates determine the specific proportionality constraints that are introduced by the multiple indicator specification, and the reasonableness of these constraints is best assessed by examining whether the proportionally constrained covariances are consistent with the actual observed covariances. For two problematic indicators, the compromise λ estimate will result in some over- and some underestimated covariances which are likely to display themselves as a telltale pattern of large and matched positive-negative standardized residuals for this pair of indicator variables.[30] With three or more problematic indicators, substantial positive and negative standardized residuals will appear, but the patterning of the signs of the residuals is not likely to follow a regular pattern.

If a multiple indicator model fails (has a large χ^2) and the diagnostics are unclear, I recommend restarting with a model containing only the single best indicator of each concept (i.e., forcing your meanings on the concepts), and then adding what are thought to be second-best and third-best indicators in separate LISREL runs. This can get tedious and expensive if there are many multiple indicators,[31] but the resulting substantive conclusions are typically worth the effort. The failure of multiple indicators is a failure of conceptualization, and this is sufficiently theoretically important to warrant publication no matter what else happens in the rest of the model.

My experience is that it is tough to get even two indicators of most concepts to cooperate, and rare to find three well-behaved indicators of a concept. Consequently, I discourage my students from using initial models with many sets of multiple indicators. All this does is overlay the diagnostics of indicator problems on top of the diagnostics of structural problems, which leads to confusion, not clarification.[32]

1.6 Retrospect and Prospect

In retrospect you should view this chapter's title, "The Constraints and Deceptions of Factor Models," as nearly self-contradictory. Part of the title is reasonable or even laudable, while another part is despicable. The *constraints* are the reasonable part. Without constraints there is no model. No estimates can be obtained and no tests can be made. All models are essentially and irremediably composed of constraints. These constraints should neither be derided nor worshipped. They are simply a component of the modeling exercise. The constraints comprising any given model, even a factor model, become laudable when the model matches data arising from a similarly constrained real world. It is the precision with which a model's constraints match the real world constraints that transforms a mere model into a trustworthy descriptive and predictive device.

The *deceptions* of the factor model, on the other hand, are insufferable. If the factor model, or indeed any model, obscures reasonable research tasks, or entangles one in ancillary practices that keep one from clearly seeing the prospects of particular research strategies, or preemptively strikes out potentially disconfirming evidence, then the model becomes the researcher's enemy.

What is being implicitly argued here is that interpreting a model should go beyond merely discussing the estimated coefficients in that model. Researchers should strive to develop a transcendent perspective on, or a transcendent interpretation of, their model by pondering and discussing the overall structure of their model and the nature of the arguments that constitute it. That is, researchers should try to characterize the single entity called "the model." Some models are bold and assertive while others are timid and non-committal, depending on the number, location, and style of the model's constraints. Other models may be characterized by the kinds of symmetries they entail, or by their funneling multiple diverse causes through a few intervening variables. By characterizing a model, one is characterizing the corresponding theory, and one thereby gains a perspective on the modeling exercise as a whole.

We must also get used to the idea that all our methodological decisions should be coordinated within this transcendent view of the model. This general stance is embodied in the admonition to *model as many of your methodological problems as you can*. Anything that unwarrantedly *influences* the data is a methodological problem. But our models are models of influences, so such problems can be represented as model segments. The incorporation of unanticipated or unplanned influences into a model eradicates any concern for the unanticipatedness of the influences. By including potentially effective methodological mess-ups into a model, one statistically controls for those mess-ups. As a bonus, one gains an assessment of the true effectiveness of the mess-ups via the estimates in that part of the model. If a mess-up is ineffective, the researcher is pardoned unequivocally. If the mess-up is effective, it has none the less been controlled and adjusted for.

One can think of methodological problems as unplanned, or unavoidable, concomitant experimental interventions. Each study's methodology is under the researcher's control, just as an experimental treatment is under the researcher's control. In this sense, learning about the effectiveness or ineffectiveness of methodological gaffes is comparable to learning about unplanned or unavoidable experimental treatments. If these accidental treatments are ineffective, one is pardoned. If they are effective, one has a fuller, though possibly serendipitously located, story about the effective sources of some modeled variable. Either way, the researcher wins.

Having broadened our appreciation of the factor model's connection to constraints, multiple indicators, concepts and the research exercise more generally, we can now consider whether estimation of a factor model should necessarily precede the estimation of any other structural equation model, as has been recently proposed.

Notes

1. Constraints are implicit in discussions of the decomposition of covariances or model implied covariances (Bollen, 1989a:34, 85; *Essentials*, 106-116, 139, 159, 163, 215-216).

2. *Essentials* (127-133) discusses maximum likelihood estimates as minimizing the discrepancies between the data covariance matrix **S** and the model implied covariance matrix Σ. The χ^2 test of model fit, discussed in Chapter 6 of *Essentials*, explores the failure of models despite the freedom to alter the coefficient estimates during the estimation process.

3. Duncan (1975:27) admonished us in this regard but this still has not become "standard operating procedure."

4. See Bollen (1989a:85) or *Essentials* (106-116).

5. Appeals to: "Isn't it obvious?", "But we all agree!", and "Are you really so dense or ill-trained that you do not understand so direct a claim?" skirt the issue of justification.

6. The only way to create looseness between "causation" and the rigidity of the proportionality implication is by claiming that this overall style of modeling is incapable of representing "true causality." It is conceivable that "true causality" might be better described by some other modeling style, but before we can develop that alternative modeling style, we will need a precise statement of the systematic discrepancies between the common jargon of "causal thinking" and the basic components of the general linear model. Here our philosopher friends can truly do us a service. But until they get around to this, you can help by attempting to verbalize the ways in which your "thinking with causes" contrasts with your structural equation model. The many parallels (e.g., one variable can cause several others, several variables can cause one variable, some causes are unknown, causes can be strung together in chains, etc.) are the ultimate justification for the commitment referred to above. It will take some "obvious" points of mismatch, encapsulated in a few memorable slogans, to drive a wedge between the jargon of causation and structural equation models.

7. The variance of y_5 would also be new data, but this additional information would be used to estimate the error variance for y_5, which is irrelevant for our current purposes.

8. Changing a λ value changes the correlation between the concept and its indicator. Since the meaning of the data has not changed, the only alternative is to admit that the change in the correlation arises because the identity of the concept (and hence its meaning) has changed. Changing the variance of the concept, even with a fixed λ, also changes the meaning of the concept because the correlation between the concept and indicator is again altered, and the meaning of the data is unchanged.

9. Values larger than 1.0 arise if $\lambda_{1,1}$ is smaller than the loadings for the other three indicators.

10. Or, in general, "are linked by $\lambda_{5,1}$ and Var(η_1) to ..."

11. Similar proportionality constraints control the covariances among the first four indicators in this model, but we are assuming these are well satisfied because of the prior success of the four-indicator factor model.

12. The following steps directly parallel the steps for the general matrix procedure (*Essentials*, 109-116) for calculating any model-implied variance-covariance matrix, and the logic grounding each step is identical. In fact, here we are merely doing piecemeal (one covariance or variance at a time) exactly what the matrix derivation does as one big step (i.e., for the full covariance matrix).

If you do not see this, return to this note after reading the steps that follow, and consider the parallel between these steps and how Eq. 3 from *Essentials* (inside front cover or page 91) is used in *Essentials* Eq. 4.25. Any row out of $\Lambda_x\xi + \delta$ (the left inner-parenthetic material on the right side of 4.25) is the equation for an X indicator. Any column out of $(\Lambda_x\xi + \delta)'$ (the right inner-parenthetic material in 4.25) is the equation for another X indicator. This provides step one. Step two is the product of these parts, namely the product of the two parenthesized sections inside the [ ] in Eq. 4.25. Step three is taking expectations (averaging across individuals) paralleling the E[ ] in 4.25. Step four notes that the left side of the equation provides covariances because these are the steps that produce covariances (*Essentials* Eq. 4.25 left side and 3.23). One then simplifies (*Essentials* Eqs. 4.26-4.29) until one has the covariances expressed in a useful form (Eq. 4.30).

13. Add E(..) around everything on the left side of the equation, and another E(..) around everything on the right side.

14. The typical simplifications are that the average or expected value of a sum equals the sum of the averages of each of the parts [E(XY + PQ) = E(XY) + E(PQ)], and factoring out constants (but not variables) from inside to outside the sum implicit in the expectation operator [E(aXY) = aE(XY) when "a" is a constant and X and Y are variables] (recall Appendix A of *Essentials*). One assumes that the means of all the variables are zero so that the formulas for the variances of, and covariance between, variables (*Essentials* Eqs. 1.6, 1.8, and 3.21) become Var(X) = E(X^2) = E(XX) (the average of the squared values of the variable) and Cov(X,Y) = E(XY) (the average of the product of the variables' values), respectively. The covariances between two variables that are assumed to be independent become zero and can be dropped from the equation (*Essentials* Eq. 1.32).

About the only complication that arises is that simplification can lead to discovery of a covariance or variance that does not exist as a fundamental coefficient in the model. This will happen whenever one of the variables used on the right-hand side of the equations beginning this process contains a variable which itself is a dependent variable in the model. If this happens, write the equations for the variable(s) that are required to construct the offending variance or covariance and use the steps summarized in the text and this footnote to reduce the problematic variance/covariance to more fundamental coefficients. Do this repeatedly if necessary. It should be obvious that calculation of a model-implied variance requires using the same equation twice to get the squaring of the variable, while calculation of a model-implied covariance requires the equations for two separate variables. Review pages 4-16 of *Essentials* before turning to Duncan (1975:6-7) if you have difficulty with this.

15. This includes multiple causal indicator models (MacCallum and Browne, 1993; Bollen and Lennox, 1991) even though we have not illustrated this style of model here.

16. Or, why repeatedly reestimate the error variance attached to mundane variables like age or income, if these can be obtained from another study that has specifically determined an appropriate error variance for the style of methodology one is using?

17. This is the exclusive mode of explanation if there is only one common factor and merely the primary component of the explanation if there are multiple correlated factors.

18. Review Figure 1.12 of *Essentials* (page 27) if this puzzles you.

19. This is the only variable whose variance and effects combine in a way that accounts for the indicator covariances without omission of the "extra" error variance term on the right of Eq. 1.41. Nor can the variable be closer to some of the items than others. Similarly biased estimates will arise whenever the true branching structure does not match with the single branch-point permitted in the factor model. If the branchings do not diverge from a common source, as in the figure below at left, then the factor model will not fit the data because locating "the factor" anywhere along the vertical chain of effects will always misrepresent the covariance implications of one or more of the error variables. The necessarily separate identity of the variables at each of the points of divergence becomes more apparent to the eye if we redraw this as on the right below.

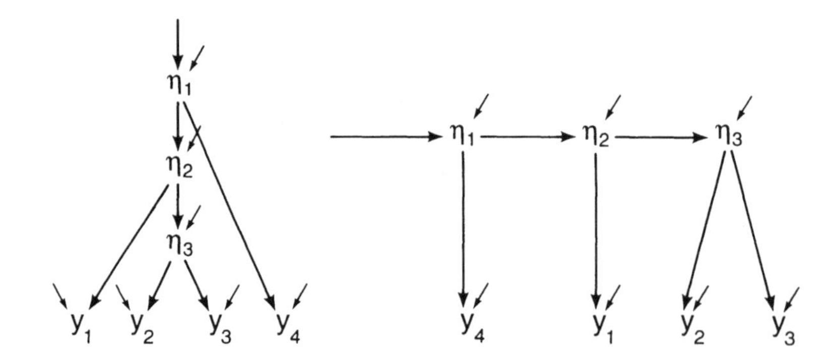

The question of whether the divergence is precisely from a single variable (rather than from several similar variables) is another way of expressing a concern for the unidimensionality of the factor. As the points of divergence are progressively squashed closer to one another (which amounts to permitting only progressively smaller error variances), one approaches a unidimensional factor.

20. There could be many common indirect causes if the effect of η_1 on η_2 in Figure 1.3A is replaced with a causal chain incorporating multiple variables.

21. Via minor changes to page 151 and by attaching the explanatory note in Figure 1.4 to *Essentials* Figure 5.7.

22. This was provided in several sections (pages 118-123, 212-218, 354-355). My relegation of multiple indicators to the later sections of *Essentials* and the absence of a strong wording for my proposed strategy fostered a misconception that I dislike multiple indicators. In fact, I admire them for their testing powers. I merely refuse to use them in a way that weakens my grasp of the theoretical meanings of the concepts in my model, as we will see below.

23. This is most easily accomplished by making a personal judgment of the percent (proportion) of the indicator's variance that is error, and then multiplying the real variance of that indicator by the specified percent (proportion) to determine the specific numeric value at which θ is fixed. The use of a percent (proportion) is intended to increase the convenience of making and reporting one's assessment, not for the purpose of standardizing. The indicators should be used with their real metrics because the model χ^2 is trustworthy only if a covariance data matrix is input. One's calculation of the metric θ value is easily checked because the complement of one's percent (proportion) of error should appear as LISREL's reported squared multiple correlation for that indicator. I have encounterd models where LISREL's generalized least squares estimator refused to closely parallel the fixed percentage of error variance, though the identical model using maximum likelihood estimation functioned flawlessly in this respect.

24. Though Joreskog and Sorbom (1993:37) illustrate the fixing of a θ variance, they do not discuss how this changes the meaning of the corresponding concept. Their claim that the "most useful and convenient way of assigning the units of measurement of the latent variables is to assume that they are standardized so that they have unit variances in the population" (Joreskog and Sorbom, 1993:173) leaves the corresponding λ's free, and hence permits variability in the indicators-concept gap. This introduces imprecision into the meanings of the concepts, even if it is conveniently introduced imprecision. The same imprecision can be found in Anderson and Gerbing's suggestion to set the metric of factors by fixing "the diagonal of the phi matrix at 1.0, giving all the factors unit variances" (1988:415) because this is essentially the same proposal.

25. Contrast this with ordinary factor analysis where one is unable to estimate, let alone test, models with two indicators per concept.

26. The distribution of the individuals' values on the concept must cover a slightly smaller range of scale scores because the variance of the concept must be less than the variance of the indicator if the variance of the concept plus some error variances is to sum to the variance of the indicator. For example, if we consider the η_1 to y_1 relation as a specific instance of preface figure column two, then the variance of the indicator is $\text{Var}(y_1) = 1.0^2\text{Var}(\eta_1) + \text{Var}(\epsilon_1)$, so η_1's variance must be lower than y_1's variance by an amount determined by $\text{Var}(\epsilon_1)$. See also the discussion in *Essentials* (119-122).

27. The multiple runs required to write this style of footnote are not too expensive if you specify no start values (NS) on LISREL's output line and enter accurate start values for all or most of the model's coefficients, based on the estimates from the model containing your preferred θ specifications. Naturally, there is almost no cost at all if you are running LISREL on a PC. *Essentials* (125) note 7 illustrates the procedure but because this note was discussing an example in which we had little substantive interest, this note is not as sharply focused on details of the meanings of the concept as I would prefer your note to be.

28. See *Essentials*, 139-140.

29. How low is low? If the best indicator is assigned five percent measurement error, and the second best indicator is thought to be about as good, yet it is estimated to have thirty percent error variance (six times as much error), I would be concerned.

30. In *Essentials* two pairs of multiple indicators were illustrated in Appendix C. These pairs were explicitly chosen because one of the pairs works well while the other pair fails. Contrast the residuals for the problematic indicators (Attend and Godexist) with the residual for the acceptable indicators (Smoking1 and Smoking4) in the last two rows and the first two columns of the residual matrix respectively (*Essentials*, 363). In this model the θ values for all the indicators had been fixed at specific values, so the huge modification indices for Attend and Godexist, and the minimal modification indices for Smoking1 and Smoking4 provide another clear indication of which of the sets of multiple indicators is contributing to the ill fit of the overall model (*Essentials*, 360).

31. The expense can be minimized by specifying start values which are very similar to the estimates obtained in the initial single-indicator run. Cost is minimized by doing two runs, one with a single indicator per concept and one with all the indicators. If this is insufficiently informative, a series of runs, using the single best indicator for all but one of the concepts, and including all the indicators for that one concept, may be required to locate the problematic set(s) of indicators.

32. MacCallum, Roznowski and Necowitz (1992) are among the many who have found that models with many multiple indicators are likely to be bad in so many ways that there is no clear path to improvement, and hence modifications are likely to end up capitalizing on chance. The strongest advice they offered was to specify multiple prior models. This advice is sound, but it is substantially off target because it fails to address the basic conceptual/meaning problem implicit if one has multiple ill-behaved indicators.

Chapter 2

Two Steps: The Factor Model before Any Other Structural Model

The proponents of a two-step approach to model estimation and development have included Burt (1973, 1976), James, Mulaik and Brett (1982), and Herting and Costner (1985). In *Essentials* (118-123), I implicitly argued against the two-step approach by defending several antithetical proposals. I hoped the two-step procedure would die a quiet death, but it was not to be.[1] Anderson and Gerbing (1988) headlined the recommendation for a two-step procedure in the *Psychological Bulletin*, Fornell and Yi (1992a) launched a counterattack, and comments were exchanged (Anderson and Gerbing, 1992; Fornell and Yi, 1992b) and the two-step procedure is still with us (Joreskog, 1993:297; Joreskog and Sorbom, 1993:113, 128).

Let me forewarn you that the issues addressed in this chapter are disorganized. There is no single straight path through this particular maze. Some seeming issues are actually tangents. Correspondingly, this chapter is structured to highlight the useful points rather than to merely determine whether one should or should not routinely adopt the two-step approach. If this is what you are seeking, let me begin by telling you that my conclusion is still to recommend against routine use of the two-step procedure, though on occasion it may be acceptable and even helpful.

Much more value can be gleaned from the glimpses the debates provide into the implicit styles of modeling adopted by the participants, and the implicit trains of thought that lead to the discussions of equivalent models, model modification, and what constitutes theory. These topics are at the limits of our current knowledge, and have not been sufficiently discussed in the literature to develop consensual understandings. But, they are fundamental and resurface in the following chapters. Some other topics that touch upon the two-step debate include scales (Section 2.1.3.1.1), indicatorless concepts (Section 2.2.3.4), and acceptable χ^2 probabilities (Section 2.3.1).

Be forewarned that many of my evaluations and recommendations are based on my views of acceptable and unacceptable research modus operandi, and few are based on purely statistical considerations. If you and I disagree on particular points, this probably stems from your accepting a general research style with which I am uncomfortable, or vice versa. I have tried to present the balanced perspective called for by Fornell and Yi (1992b) but I have resisted tentativeness and hedging. If I have constructed this chapter adequately, you should sense me stomping around near the limits of your modeling foundations. Naturally, I anticipate very different reactions, depending on whether my stomps land on, rather than in front of, your modeling toes. Cathartic marginal notes are certainly in the spirit of this chapter.

Let us begin by summarizing the basics of the two-step proposal. *Step one* involves estimation of a so-called *measurement model*, which is usually a confirmatory factor model allowing the concepts to freely covary with one another while attention is focused on which concepts influence which of the observed indicators. This is typically implemented by including all the indicators, both those that will ultimately be called x's and y's, as x variables and then estimating only the x-measurement model in LISREL, with Φ free, θ variances free, and selected λ's free.

In *step two* the measurement model obtained from step one is used as a component of a full LISREL model incorporating the theoretically postulated *structural effects among the conceptual variables*. The covariances among the concepts from step one are largely replaced with directed β and γ effects in step two.[2] There is some variation in the degree to which the λ's (factor loadings) from step one are incorporated in the second-step structural model. The locations of the λ's are typically maintained, but these coefficients may or may not be reestimated (Fornell and Yi, 1992a).

2.1 A Recent Exchange

This chapter is structured to generally follow the critique of the two-step procedure offered by Fornell and Yi (1992a). This is a useful place to begin because they raised some new issues in addition to condensing the basic points made by Anderson and Gerbing (1988). The critique by Fornell and Yi (1992a) was followed by comments from Anderson and Gerbing (1992), and a rebuttal by Fornell and Yi (1992b). The points made in these works are melded together in the discussion of the four questionable assumptions of the two-step procedure that are raised by Fornell and Yi (1992a).

Fornell and Yi (1992a) attribute four putative assumptions (assertions) to the two-step procedure, each of which they challenge in the course of providing their critique. These assumptions are not worth memorizing. We will end up discarding some and modifying others, and adding several of our own.

Fornell and Yi question the two-step, factor-model before structural-model procedure because *in their view, this unjustifiably assumes that*:

1. Theory and measurement are independent of one another or can be treated as such.
2. Measurement validity established in the first step (via confirmatory factor analysis or separate factor analysis) can be generalized to other model specifications.
3. The estimators of a two-step approach are (asymptotically) unbiased, consistent, and efficient.
4. The statistical test done during step one is independent of the test done at the second step.

(Fornell and Yi, 1992a:295)

We will examine each of these assertions in turn.

2.1.1 Assertion 1

The general thrust of the first assertion is a claim that the first-step factor model is fundamentally concerned with measurement while the second-step structural model is theoretical. Fornell and Yi (1992a) question whether we can, or should, routinely separate measurement from theory, and hence they defend the view that measurement issues cannot routinely be decided without implementing theory.

The first prong of Fornell and Yi's (1992a:296) attack on the idea that the structural portion of the model (the second theory-implementing step) can be prevented from influencing the measurement properties, is to point to the seeming inseparability of measurement from theory in the natural sciences and assert a parallel inseparability in the social sciences: "The *interpretation* of an observation is always done in the context of some theoretical framework"; "Theory not only serves as a basis of interpretation but also determines what is to be counted as an observation"; "If we acknowledge that our measures are theory-laden (in one way or another), we may begin to understand why Theory X produces different measurement properties (e.g., loadings) than Theory Y" (Fornell and Yi, 1992a:296-297). The second prong of Fornell and Yi's attack is an example showing how λ loadings can change if one changes the structural model.

2.1.1.1 Reply to Data and Measurement as Theory Laden

In defending the two-step procedure, Anderson and Gerbing do not rise to the bait of separating measurement from theory. They agree that observations or data are theory laden (1992:322). They see the researcher's specification of the number of concepts, and the assignment of particular indicators as loading on particular concepts in the factor model, as being based in theory. Hence, Anderson and Gerbing agree with Fornell and Yi in viewing Assertion 1 as false, but this does not dampen Anderson and Gerbing's support for the two-step process. To them, this merely renders the assertion nondescriptive of the two-step procedure, and hence irrelevant.

But the attestations of all concerned to the general principle that measures are theory laden masks continuing disagreements over the details of how theory influences measurement, and what this implies about standard operating procedures. Anderson and Gerbing seem to sense this, for they go on to reconfirm their acceptance of epistemological fallibility, and the hypothetical, imaginative, nature of constructs. In Chapter 1 we argued that not only are *concepts* theoretical fictions, so are *effects*. The effects of concepts on indicators are no less imaginative and fallible than are the effects of concepts on other concepts. Indeed, λ effects might even be viewed as *more* mysterious or hypothetical than β or γ effects because they are fictions that straddle the real and fictional worlds, rather than merely providing fictional coordinations between fictional entities (concepts). If we start from a position that all fictions are created equal, how can we recover a claim that one of these classes of fictions (the now theory-laden measurement fictions) are to be treated preferentially by being examined first in the two-step process?

Pointing to the separate diagnostics accompanying the models at each of the two steps as providing superior diagnostics over the single set of diagnostics in the one-step approach (Anderson and Gerbing 1992:323) begs the question. There are many two-step approaches that can be adopted. Any pair of nested models can claim to constitute a two-step procedure (the model with the additional constraints being the second step that is nested within the less constrained first step). The researcher has two outputs to scan and seemingly twice as much diagnostic information. If we grant that the reasons for choosing the two nested models should match the substantive points at issue in the literature addressed by the model, we are proposing that any of a host of alternative nested models are preferable to any single model. Any pair of nested models provide "separate assessments" of two model components and hence any of the multiple possible styles of nested models can claim to have "superior diagnostics over a one-step approach as to the specific sources of ... fallibility" (Anderson and Gerbing, 1992:323).

What then makes Anderson and Gerbing believe that their particular style of nesting (model subsumption) warrants routine preference, in contrast to allowing substantive issues to determine the style of nesting to be employed? Their justification here touches the central issue. "Because the constructs are allowed to freely intercorrelate in a measurement model, lack of fit must be due to fallibility in a researcher's theory of how one or more of the measures are related to the constructs" (Anderson and Gerbing, 1992:323).

What Anderson and Gerbing are alluding to here is that any particular set of numerical values for β, γ, ϕ, and ψ imply a particular set of covariances among the concepts.[3] The covariances among the concepts are freely estimated during the first step of the two-step procedure, and hence this allows the best of all the possible concept covariance matrices to be estimated. There is no way for a structural model with specified directed effects to improve upon the step-one model's fit to the observed indicator covariances.[4] Directed effects exactly implying (reproducing) the step-one concepts' covariances only do as well at reproducing the indicator covariances as did the step-one model. Directed effects unable to provide this covariance matrix do worse, since any conceptual variable covariance matrix other than that estimated in step one must fit the data less well. It is this seeming "hands off the structural part of the theory," or "having let the structural part do the best it ever could" that is appealing. This is the central justification that Fornell and Yi are challenging and which we chip away at as we continue our investigation of the remaining assertions.

It should be clear that the separation of structural (β and γ) effects from measurement (λ) effects is central to the issue. This is implicated in the limited way Anderson and Gerbing grant the intrusion of theory into measurement. Their acceptance of data as being theory laden specifically avoids saying anything about β or γ *altering* measurement, and confines theory's connection to the data to counting the concepts, "identifying" the concepts, and postulating links between the concepts and indicators. So the agreement between Anderson and Gerbing, and Fornell and Yi, that data are theory laden glosses over continuing fundamental disagreements. Justifications for the relevance of β and γ effects to measurement decisions appear in Section 2.1.3.2 on sparse models, Section 2.2 on the implicit wider context for this debate, and in our discussion of model modification, which we do not get to until Chapter 4. Thus the issue is not whether theory influences measurement but how theory influences measurement.

2.1.1.2 Reply to the Example

You will recall that the second prong of Fornell and Yi's (1992a) attack on Assertion 1 was an example displaying a phenomenon which should be disturbing to proponents of the two-step approach, namely that λ estimates can change if one changes the structural model. Fornell and Yi begin by estimating the model from Anderson and Narus (1984) illustrated at the top of Figure 2.1. They then posit an alternative theory which interchanges the Control and Satisfaction concepts, and reestimate the model. These two models fit the data about equally well (as judged by χ^2) though the β and γ values differ markedly. The point of interest to Fornell and Yi, however, is not the effects among the concepts but the estimates of the λ values linking OCL to its two indicators. In the original model, the λ for the first indicator is about double the λ for the second indicator. In the revised model the relative magnitudes are reversed, with the second indicator's λ being slightly larger than the first indicator's λ.

The challenge is clear. Given essentially equivalent fit of the two models, the *measurement* connections between OCL and its indicators seem to be dependent on the *structural effects among the concepts*; namely whether OCL influences Satisfaction or Control.

Anderson and Gerbing (1992:324) dismiss the OCL example as illustrating "the known instability and problematic nature of estimated constructs defined by two indicators" as discussed by Gerbing and Anderson (1985) and point out that the loadings for the concepts having more indicators were much more stable. (Satisfaction has five indicators, OCLA four, and Control three.)

This reply is ineffective for two reasons. First, the Gerbing and Anderson (1985) paper addresses fluctuations in sample estimates arising from repeated random samplings from a population, but the Fornell and Yi example shows differences in estimates associated with using different structural models *within a single sample*. There simply is no sampling variability to provide any explanation of the differences in λ values observed by Fornell and Yi.

Second, any reply appealing to the special nature of models with two indicators implicitly admits that the two-step procedure is inappropriate for models incorporating concepts with two or fewer[5] indicators. Pointing to *any* reason as explaining why models with two indicators behave unacceptably when used in the two-step procedure specifically acknowledges the unacceptable behavior of the procedure for those models, and hence grants that *the procedure is inappropriate for use with models including even one concept having fewer than three indicators*. This excludes most models, or at least models with single indicators for age, sex, education, etc., and models following the measurement specification procedures suggested in Chapter 1 and in *Essentials* (118-123).

Figure 2.1 Lambda Estimates as Structural Model Dependent.

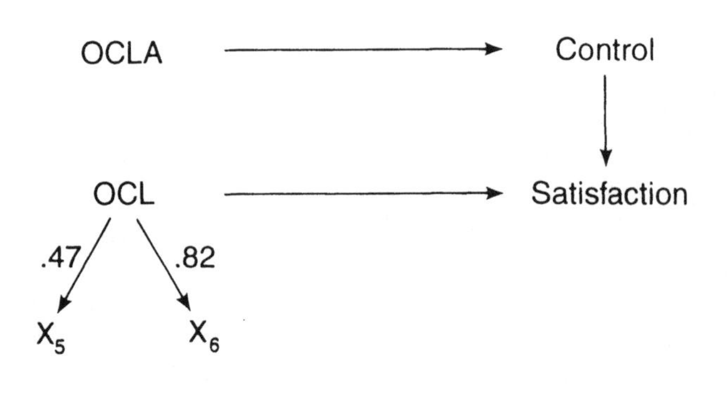

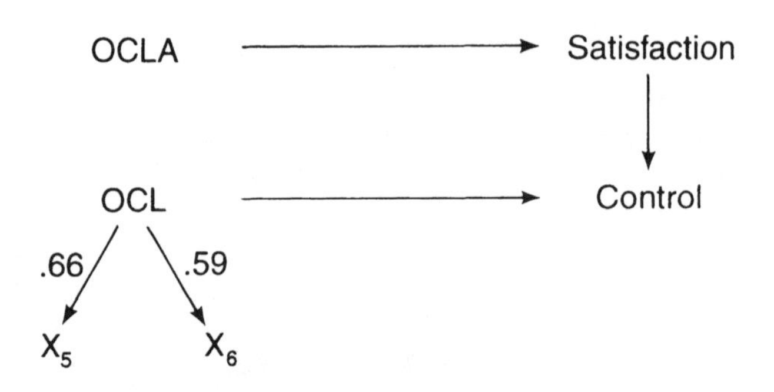

OCL: Outcome given Comparison Level
OCLA: Outcome given Comparison Level for Alternatives
Control: Manufacturer Control
Satisfaction: Cooperation/Satisfaction

Note: Error terms, four indicators of OCLA, three indicators of Control, and five indicators of Satisfaction are not illustrated since they are irrelevant to the current discussion.

2.1.2 Assertion 2

The second assertion by Fornell and Yi (1992a:295), namely that a two-step approach unjustifiably assumes "*measurement validity established in the first step (via confirmatory factor analysis or separate factor analysis) can be generalized to other model specifications*," only weakly foreshadows the nature of the substantive challenges they subsume under this heading. Fornell and Yi's two major challenges focus on equivalent structural models, and on separate factor analyses.[6]

2.1.2.1 Equivalent Models

The equivalence of models becomes an issue when Fornell and Yi construct three models of the possible connections between the four concepts in Ajzen and Fishbein's (1980) model. We will have no reason to refer to the substantive meaning of these concepts so we refer to these as η_1 through η_4 in Figure 2.2. The model in part A of this figure depicts the saturated correlation model that arises when all the correlations among the concepts are freed during the first of the two steps. Model B is a saturated recursive model that postulates two of the four concepts as dependent while the remaining two are correlated exogenous variables.[7] Model C is like B but it is not saturated because two of the effects are omitted since they are postulated to be zero. Fornell and Yi are correct in arguing that models A and B are equivalent (their appendix shows the correspondence between the models' parameters, and see Chapter 4 below), but they are incorrect in the conclusion they draw from this. They suggest that if one's true model is C, and step one of the two-step process uses model A (which is equivalent to the "wrong" model B) in evaluating the indicators of these concepts, this "yields an evaluation of measures in the wrong context" (Fornell and Yi, 1992a:305). Model C, not B, is the proper theoretical context for evaluation of the measurement structure.

Anderson and Gerbing (1992:325) correctly point out that this argument is flawed because there is no way to argue which particular one of the multiple equivalent saturated recursive models is "the one" underlying the correlational model. Fornell and Yi selected the saturated model (our model B) that is most similar to the model of substantive interest (model C), but the correlational model A bears neither more nor less commitment to this particular directed and saturated model than to any of the multitude of other saturated models (including models with reversed effects). Hence, while Fornell and Yi's demonstration of the equivalence of models A and B is correct, model A is also equivalent to all

Figure 2.2 Some Nested and Equivalent Models.

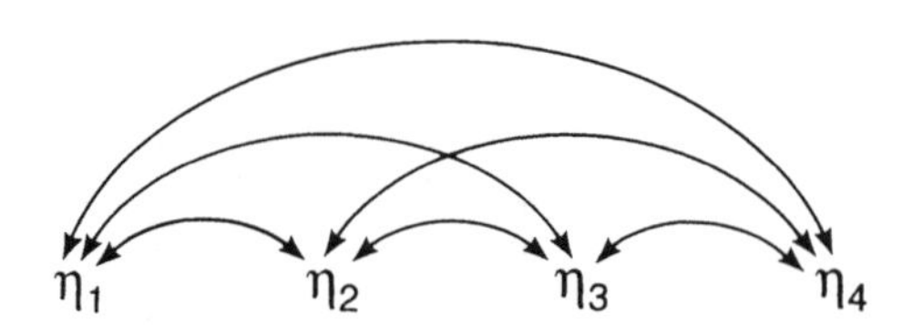

A: A saturated correlation model.

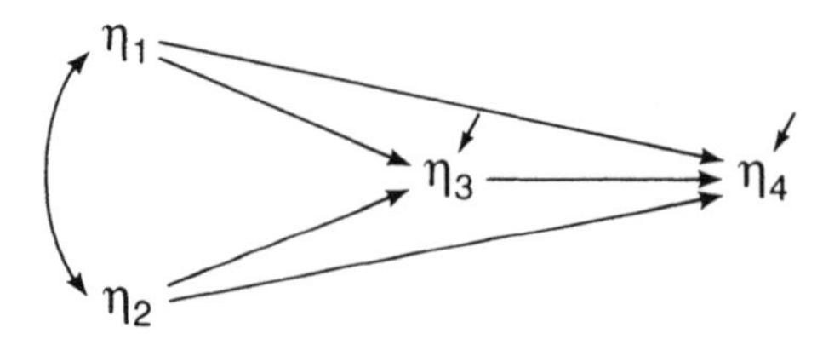

B: A saturated effect model.

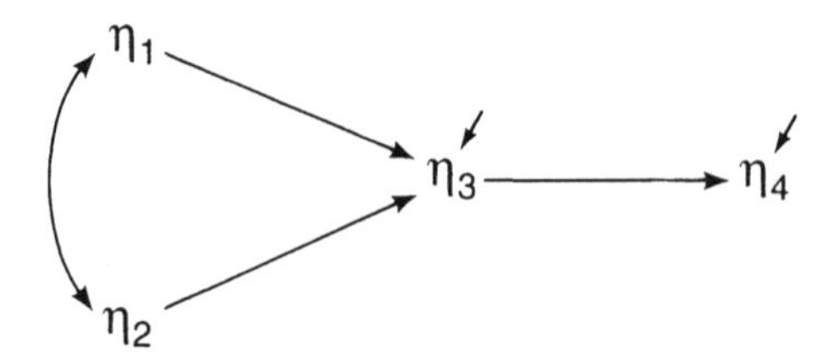

C: An unsaturated effect model.

the models that can be created by randomly shuffling the η's among the various locations in model B.[8] Hence, there is no particular "wrong" directed model used in the first step. As a result, this particular challenge is ill conceived, and is of no assistance in deciding to, or not to, recommend the two-step procedure.[9] The quest for the ways in which a correlational representation can be a wrong representation of a directional model harkens back to the fundamental issue raised near the end of Section 2.1.1.1.

2.1.2.2 Separate Factor Analyses

Fornell and Yi's second attack on Assertion 2 is instructive but equally unfounded. They begin by creating a hypothetical model with known parameters. The model resembles our Figure 1.2C, with concept η_1 causing η_2, but each of the concepts has four indicators. Fornell and Yi then use the known model and prespecified coefficients to obtain the covariance matrix for the indicators,[10] and they use this covariance matrix as data for estimating three other models. Two of these three models are separate factor analyses in that one concept and four indicators comprise each model. The instructive twist on these models is that one of the four selected indicators in each of these models was used as an indicator of the wrong concept. Imagine y_5 as replacing y_4 in our Figure 1.2C model. The other model corresponds to our y_4, replacing one of η_2's true indicators.

These two separate factor analysis models both perfectly fit the data to which they are applied (the covariances among the appropriate four indicators). The reason for these perfect fits is clear. Look at Figure 1.2C and note that y_5 is indeed an indicator of η_1 because η_1 causes y_5. The effect is indirect (but then the effects to y_1, y_2, or y_3 may also be indirect because they may include implicit intervening variables) and the model fits because similar proportionality constraints are implied for y_5's covariances in our models 1.2A and 1.2C.[11]

Fornell and Yi's third model incorporates both the concepts (with η_1 causing η_2) and the sets of four indicators containing the misplaced indicators. This model fails. Fornell and Yi conclude that even though the structural part of the theory is correct (η_1 was allowed to influence η_2, as in the true model), this was not detected by the two-step approach using separate factor analyses as the first step. The perfect fit of the separate factor analyses suggests that the measurement was perfect (which it is not), while the failure of the full model incorporating the directed effects at the second step seems to say the structural model was in error (which it is not).

Fornell and Yi's demonstration stands as a clear warning to those doing separate factor analyses that trouble awaits any attempt to combine separate analyses into more comprehensive models. The double-proportionality constraints discussed in Section 1.3.4 simply will not hold if the comprehensive model includes both the true causal factor (concept) and the putative causal factor (concept) for a misplaced indicator.

As sound as this demonstration is, and as earnest the warning to factor analysis devotees, this demonstration fails to challenge the two-step procedure favored by Anderson and Gerbing. Their proposed first step is not two separate factor analyses, but a factor analysis incorporating both η_1 and η_2 with these factors covarying, instead of affecting one another. This joint factor analysis fails just as dismally as the last of Fornell and Yi's models because the double-proportionality constraints are the same whether η_1 causes η_2 or whether η_1 correlates with η_2.[12] Hence, Anderson and Gerbing's step-one model would fail and they would conclude that there are measurement problems, as indeed there are.

The seemingly devastating absence of problems in the measurement step (despite real problems) and the presence of problems at the structural step (when in fact there are none) are artifacts created by the separateness of the factor analyses Fornell and Yi use as the first step in their illustration. The separateness of the analyses means that the indicator covariances crossing between the two sets of items (between the two concepts) are not utilized in the first step of the model. It is precisely these covariances that carry the bad news, and it is to Gerbing and Anderson's credit that their proposed first step incorporates this information.

We will have more to say about separate analyses in our discussion of Assertion 3. For now, however, we must conclude that Fornell and Yi have failed to make a case against Assertion 2, though they have convincingly demonstrated that separate factor analyses cannot be accepted as constituting the first of the two steps.

2.1.3 Assertion 3

The third assumption Fornell and Yi (1992a) propose as unjustifiably underlying the two-step approach is that *the estimators of a two-step approach are (asymptotically) unbiased, consistent, and efficient*. The terms unbiased, consistent and efficient refer to desirable statistical properties for estimates,[13] but it is unclear as to whether the "the" in the assertion refers to the estimates produced at the first step only, the second step only, or both steps. Fornell and Yi also provide two attacks on this assertion: the potentially biased nature of separate factor analyses, and the estimation of irrelevant parameters.[14]

2.1.3.1 Bias in Separate Factor Analyses

We continue to use the phrase "separate factor analyses" to refer to doing a separate factor analysis for each concept in one's model as the first, or measurement, step prior to integrating the factors (concepts) into a structural second-step model. "Confirmatory factor analysis" refers to a first-step factor analysis simultaneously including all the concepts, and estimating the correlations among those concepts along with selected factor loadings (λ's).

How can separate factor analyses lead to biased estimates? The top portion of Figure 2.3 depicts a world in which two concepts (η_1 and η_2) each have three indicators in addition to sharing one indicator between them. The bottom half of Figure 2.3 focuses on the concept η_1 and asks us to imagine the asterisked loadings that would appear if a separate factor analysis was done for η_1 alone. That is, the separate factor analysis (bottom figure) ignores the presence of η_2 in the real world (top figure).

We can write two equations describing the structural sources for the shared indicator y_3. The equation for y_3 in the real world (Figure 2.3A) is

$$y_3 = \lambda_3\eta_1 + \lambda_8\eta_2 + \epsilon_3 \qquad 2.1$$

while the equation for y_3 in the model the researcher estimates during a separate factor analysis is

$$y_3 = \lambda_3^* \eta_1 + \epsilon_3^*. \qquad 2.2$$

We can get the covariance between η_1 and y_3 (the covariance between the concept and one of its indicators) by multiplying each of these equations by η_1 and taking expectations (i.e., using Duncan's procedure as discussed in Section 1.3.3). Assuming the independence of the error variables and that the variables are recorded as deviations from their means, we find, respectively,

$$\text{Cov}(\eta_1, y_3) = \lambda_3\text{Var}(\eta_1) + \lambda_8\text{Cov}(\eta_1,\eta_2) \qquad 2.3$$

and

$$\text{Cov}(\eta_1, y_3) = \lambda_3^* \text{Var}(\eta_1). \qquad 2.4$$

The left sides of these equations should be equal (we are not allowed to change our conceptualization of η_1 and the data does not change); hence the right sides should also be equal:

$$\lambda_3^* \text{Var}(\eta_1) = \lambda_3\text{Var}(\eta_1) + \lambda_8\text{Cov}(\eta_1,\eta_2). \qquad 2.5$$

Dividing both sides by $\text{Var}(\eta_1)$ provides

$$\lambda_3^* = \lambda_3 + \lambda_8\text{Cov}(\eta_1,\eta_2)/\text{Var}(\eta_1). \qquad 2.6$$

This informs us that the estimate of the λ_3^* loading of y_3 on η_1 which the researcher observes in the separate factor analysis differs from the true loading λ_3 by an amount which depends on the three terms on the right: the strength and sign of the relationship between the concepts ($\text{Cov}(\eta_1,\eta_2)$), the

Figure 2.3 Bias in Separate Factor Analyses.

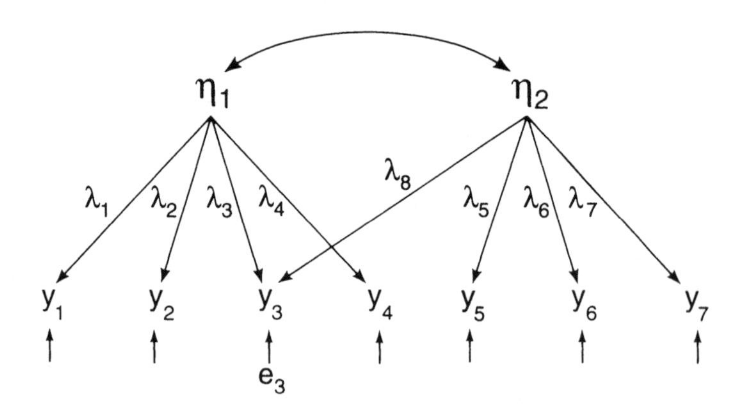

A: The true causal structure.

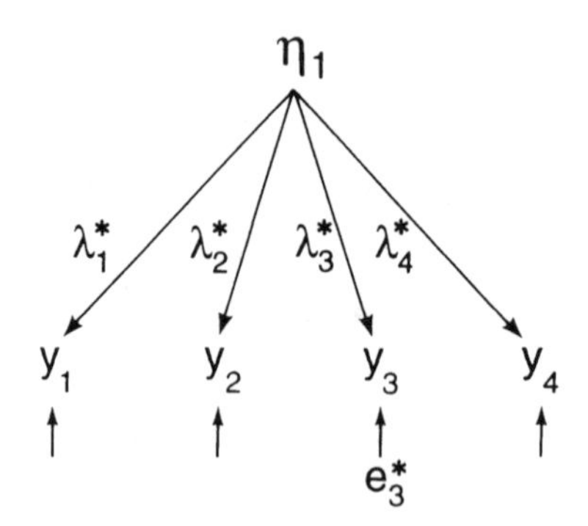

B: A separate factor analysis.

strength and sign of the influence the excluded η_2 concept has on y_3 (namely λ_8), and the variance of the concept we are focusing on ($\mathrm{Var}(\eta_1)$). The loading estimated in the separate factor analysis may be too high or too low (depending on the signs of λ_8 and $\mathrm{Cov}(\eta_1,\eta_2)$) and may be markedly or marginally different from the true value (depending on the magnitude of λ_8, $\mathrm{Cov}(\eta_1,\eta_2)$ and $\mathrm{Var}(\eta_1)$). Hence, separate factor analysis can lead one astray by providing biased estimates if a separate analysis ignores correlated concepts that are slated for later inclusion in the model and that influence some of the items.[15]

In the context of the two-step debate, this means that if the estimates obtained from the separate first step are used as *fixed* coefficients in the structural model, biased estimates may be being inserted into the second-step model, which is clearly unacceptable. Furthermore, moving from a separate factor analysis first step, to a second step including directed effects among the concepts η_1 and η_2, and either *fixed or free* λ values, leads to a model that has precisely the problem the two-step procedure is supposed to eliminate. With η_1 and η_2 in the model, the diagnostics will call for a sharing of indicator y_3 and the diagnostics for measurement are again mixed with the diagnostics for the structural/conceptual part of the model.

The bias in the estimates and the diagnostics calling for measurement (λ) revisions at the second step could have been avoided by doing a confirmatory factor analysis incorporating both η_1 and η_2. The sharing of the y_3 indicator would have been found at step one, since the estimated factor model would correspond to the true model (Figure 2.3A).

The challenge provided by bias and/or the remixing of measurement and structural diagnostics resulting from separate factor analyses might be dismissed as being irrelevant to the majority of applications of the two-step procedure because separate factor analysis is employed much less frequently than first-step confirmatory factor analysis. Furthermore, the bias is diluted if one reestimates the λ coefficients at step two rather than entering these as fixed values (though the measurement and structural diagnostics remain confounded). It is presumably for these reasons that Anderson and Gerbing (1992) do not even bother providing any response to this particular criticism by Fornell and Yi.

But we would be missing the import of the potential biases in separate factor analyses if we confine our view to current users of the two-step procedure. Separate factor analyses have been the operating norm within psychology and education for decades. Separate factor analyses purporting to demonstrate the soundness of measures have routinely been justified by claiming that the authors are doing the measurement step one so that other researchers will finally be able to use the new scale or measure, to get on with the admittedly important structural step two. The rhetorical question, "How could one possibly determine the causes and consequence of concept *X* unless one can measure *X*?" begs for the answer, "OK. Let's get the measurement issue settled *first*. *Then* we can get on with determining the structural world in which X is imbedded." Hence, while this challenge is not particularly devastating to the current two-step literature, it stands as a likely impediment to anyone implicitly using the two-step by employing indicators or scales "justified" by lengthy traditions of separate factor analyses.

Alas, there is a rug this particular problem may be swept under. Researchers can skirt this particular issue by employing "the *scale* measuring the concept" rather than testing the veracity of the common factor conceptualization by entering several of the best items as multiple indicators in the second-step structural model. Using a single scale reduces the interbattery covariances highlighted in Figure 1.2E to a single row (or column). This eliminates the need for between-row proportionality constraints as discussed in Section 1.3. The single remaining row of covariances for the scale can be relatively easily accounted for (modeled) by direct effects among the concepts. As a consequence, one robs oneself, and the literature, of the possibility of disconfirming the unidimensionality of the proposed scale or measure by employing that scale or measure rather than several of the best scale items in the second-step structural model.[16]

But there is some natural justice in this particular part of the modeling world. Besides having to bite their tongues as they now revise their epistemological allegiances by using/defending single indicators of concepts, researchers inclined to hide this problem by using scales "justified" by factor analysis will end up using single scales that are in fact composites of multiple dimensions. The mixing of multiple component dimensions in the single indicator variable is likely to result in the indicator displaying rather complex associations with other concepts, which in turn is likely to hinder attempts to place the indicator in any simple structural theory. Even if the researcher has managed to guess the proper structural model, he/she is testing the model by improperly connecting the concepts to the scale indicators (single variables created from multiple components). The result is likely to be a mess. Those inclined to ignore this particular problem are reducing their own chances of ever finding a clean and simple model. Even if they are lucky enough to have initially postulated the true and elegantly simple structural model, the mixing of the concepts within the indicators will mitigate against any clear confirmation of the structural model.

2.1.3.1.1 Scales as Trouble Makers

The devious nature of scales deserves a closer examination. Imagine I make a scale for Concept-X by calculating a weighted sum of three items, such as

$$\text{Scale-}X = 1.0\text{Item-1} + 0.5\text{Item-2} + 2.0\text{Item-3}. \qquad 2.7$$

Now, what are the causes of the Scale-*X* scores? Hidden within the "I make" near the beginning of this paragraph is the precise and complete causal structure. I, as researcher, caused the computer to use just these three items, and I caused it to weight them 1.0, 0.5 and 2.0 in producing Scale-*X*. Furthermore, I causally accomplished this without error. There is no "e" at the end of Eq. 2.7. Indeed, the causal world corresponding "my making Scale-*X*" or Eq. 2.7 can be depicted as Figure 2.4A.

You should now be tempted to point out that this is the wrong question. Should we not be concerned with the causes of Concept-*X* rather than the causes of Scale-*X*? After all, Scale-*X* is merely being introduced as an indicator of Concept-*X*. This is the story depicted in Figure 2.4B. But this particular story does not provide even the whole story, let alone the whole truth. As things stand, there seems to be a contradiction between this story and Figure 2.4A, where the items, and not a fictional concept, cause Scale-*X*. The whole story (Figure 2.4C) goes something like this. The Concept-*X* is thought to contribute to each of the items, and I created a weighted sum of the items in the hopes of tidily repackaging these back into something that reflects the *X*-ness of the concept.[17]

Note that a factor model (the dashed box in Figure 2.4C) has re-emerged at the very core of this justification. The scale for *X* remains on the right of the model, the interesting causes on the left, and the part of the story that was being hidden by the locutionary curtness above was our sneaky old friend the factor model!

Now for the troubling complexity. What if the factor model is inappropriate in that one of the background variables causes one of the indicators directly as opposed to, or in addition to, being channeled through Concept-*X*? This questions the very heart of the conceptualization, for it challenges whether Concept-*X* really does absorb the effects of disparate sources, and then proportionately redistributes these effects among the items. This will not have been checked out in any prior factor analyses since no indicators of the causes of Concept-*X* (i.e., no indicators of other concepts) will have been included in any of the initial separate factor analyses.[18] Nor will this show up if the model in Figure 2.4B is analyzed. The diagnostics calling for an effect from the background cause to the Scale-*X* (which contains the unruly indicator) will be equivocal. Most of the items contained within the Scale-*X* do not require the troublemaking direct effect from the background concept, and hence the offending item is likely to have its contribution to the diagnostics diluted. Consequently, the diagnostics focused on a direct contribution of the background concept to the Scale-*X* become equivocal and unpredictable.[19]

Without the inclusion of the items in the model, the core of the argument, namely the ability of Concept-*X* to absorb disparity and emit proportionally, goes unexamined and untested. Hence, one must either include

Figure 2.4 The Justification for Scale Construction.

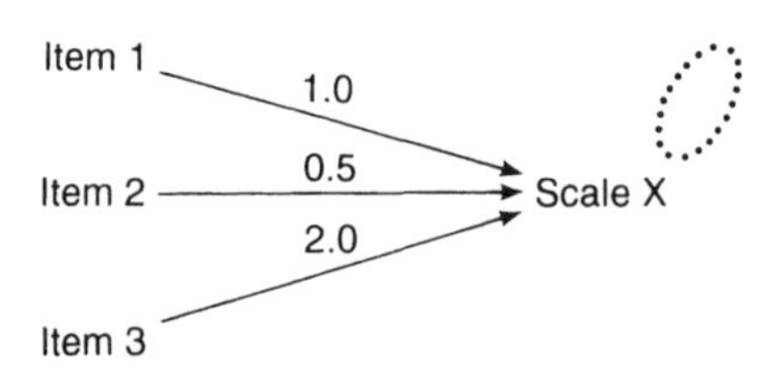

A: Creating a scale.

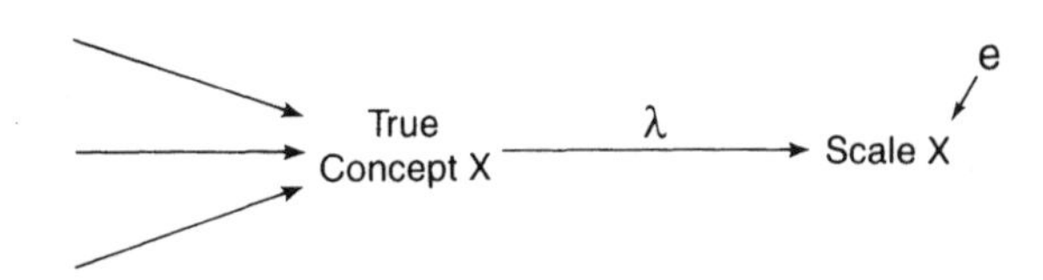

B: The scale as an indicator.

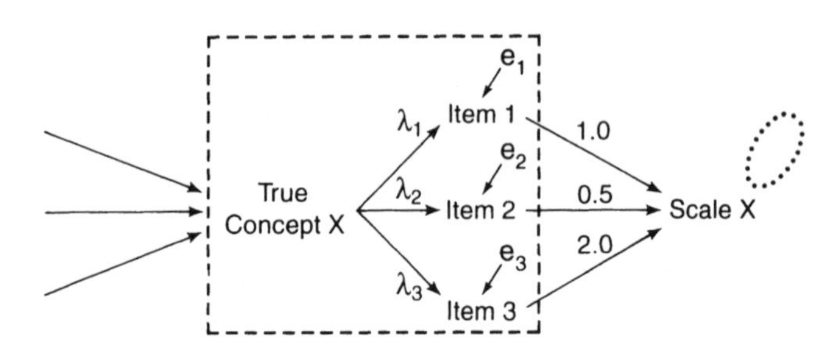

C: The implicit factor model.

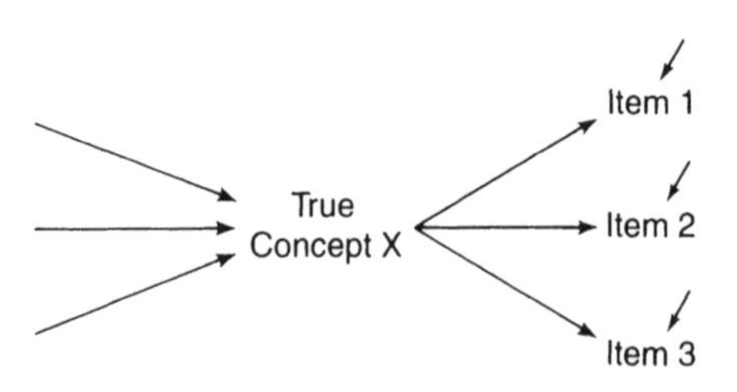

D: A step-two model reemerges.

the items (as in Figure 2.4C) or be guilty of neglecting an obvious test of the core of one's conceptualization. But if one is going to include the items, the scale becomes redundant, so we might as well work with the model in Figure 2.4D, in which case we are back to defending a single-step combined measurement and structural model.

2.1.3.2 Irrelevant Parameters

We return now to Fornell and Yi's second argument against Assertion 3. This argument concerns irrelevant parameters and it considers what happens if the true structural model is sparse, in that it contains few effect coefficients to be estimated, but one does a first-step confirmatory factor analysis that is saturated, in that all the covariances among all of the concepts are estimated. Here we are imagining a model with few structural coefficients as the true model, and estimating a model with many structural coefficients at the initial step.

Fornell and Yi (1992a:326) point out that estimating unnecessary parameters in both single-equation regression models and nested structural models (Bentler and Mooijaart, 1989) leads to standard error estimates that are larger than the standard error estimates in the more parsimonious model.[20] Anderson and Gerbing (1992:328) reply that while this is true, the magnitudes of the increases in standard errors are very small, and are likely to be smaller than biases introduced by misspecifications created when unjustifiably sparse models are estimated. It is, after all, known that misspecified models are prone to providing biased estimates (recall the discussion of bias in Section 2.1.3.1, or see Bollen, 1989a). Because of their very sparseness, sparse models are more likely to be misspecified, and the bias thus created is likely to be more serious than minor changes in efficiency due to the increase in standard errors accompanying the estimation of a few extra and unnecessary coefficients. So Anderson and Gerbing have again successfully defended the two-step procedure from Fornell and Yi's jabs.[21]

But a major point is being missed by both sides. To me, the presaging of a sparse structural model with a saturated first-step model should raise substantial concerns for the possible underidentification of the step-one measurement model. Consider a just identified model that combines a sparse structural model with a complex measurement model. That is, a model in which the sparseness in the structural part of the model is used to assist in estimation of a measurement model that would otherwise be unestimable (underidentified) because of the high density of its coefficients. The additional coefficients required when estimating all the covariances among all the concepts during the first-step measurement model would make the first-step model unidentified and hence unestimable.[22]

It is difficult to say how many of the models appearing in the literature fit this scenario, but no matter how few there currently are, we should be actively seeking such models. This is a style of model where a sparse theory lets us do something we otherwise could not do, namely get estimates of the full model. This epitomizes the potential contributions of theory to the research process. Instead of being squandered on the scientific esthetic of parsimony, the sparseness of the structural model is being put to a substantive research use. It lets one estimate an inordinately complex measurement structure that could not be estimated without the aid of the parsimonious theory. If demanding a two-step approach keeps even one researcher from utilizing a sparse theory in this truly helpful way, the two-step should be chastised for academic stifling![23]

This discussion harkens back to the issue of whether β and γ effects are part of the theory that makes data theory laden (Section 2.1.1.1). The theoretical sparseness required to allow a sparse theory to assist with estimating an inordinately complex measurement model is unlikely to arise from an entirely correlational model. There is no way to justify the omission of some of the concepts' correlations without appealing to the true underlying causal process that is producing those correlations, and this appeal necessarily invokes directed effects. Here then, we find a reason to claim that β and γ coefficients must be part of the theory that can make data theory laden.

This also provides a critique of the view that allowing the covariances among the concepts to be free keeps one's structural-theoretical hands off the data. Whenever a sparse structural model assists estimation of a dense measurement model, freeing all the concepts' covariances in step one amounts to the laying on of meddlesome hands, not keeping one's hands off!

2.1.3.3 Summary and Overview of Assertion 3

Neither of the critiques of Assertion 3 proffered by Fornell and Yi is persuasive. The first critique (bias in separate factor analyses) simply does not describe current use of the two-step procedure (confirmatory factor analyses seem to be the norm), and the second (increasing standard errors if irrelevant coefficients are included in an equation) is of relatively minor importance and is opposed by an equally valid counter-argument (potential bias caused by unjustifiably excessive parsimony).

Taking a wider perspective, however, the potential for bias in separate factor analyses has important implications. Psychology and education are likely to spend the next decade undoing the biases introduced by decades of commitment to separate factor analyses. This undoing is likely to take two basic forms: a) messy structural models being required when scales based on separate factor analyses are used, and b) ill-fitting models with measurement problems when multiple indicators are selected on the basis of prior separate factor analyses.

I have a dream that someday researchers will routinely attempt to use LISREL to do things that are otherwise impossible. Using sparse theories to assist in estimation of otherwise underidentified measurement models is a scene from that dream. The two-step procedure stands between us and the realization of this dream.

2.1.4 Assertion 4

Fornell and Yi's final proposed "unjustifiable assumption" of the two-step procedure, namely that *the statistical test in one step is independent of the test in the other*, sounds so dryly statistical that its resolution should be relatively uncomplicated, but it is not. The issue is whether or not the χ^2 test for the step-one measurement model is independent of the χ^2-difference test constructed as the difference between the χ^2's for the step-one and step-two models, and the difference between their degrees of freedom.[24] If there have been no data-prompted revisions to the models (e.g., the tests are preplanned) and if the models are truly nested, then the tests are independent, as proven by Steiger, Shapiro and Browne (1985).[25]

The Steiger, Shapiro and Browne (1985) proof is actually mute as to how the nested models came to acquire their current specification. Anderson and Gerbing feel this silence entitles them to conclude that "whether or not respecification has been employed, the adequacy of the researcher's structural submodel can be assessed independently of the fit of the measurement model" (1992:330). In contrast Fornell and Yi maintain that "once one starts adjusting a model in the light of data (here, measurement part), the model loses its status as a hypothesis" (1992a:313). That is, if model revisions have been prompted by a failing first step, we have entered an exploratory rather than confirmatory mode of analysis and hence the second test loses its independence from the first test.[26]

Let me construct a scenario that supports Fornell and Yi. Imagine I estimate a step-one model. It fits admirably, so I make no modifications to my measurement model but I notice that all but one of the correlations among the concepts are estimated at values of zero. On this basis I modify the second-step structural model. I discard all my prior theory, and now postulate all the effects to be zero except for a single direct effect between the concepts whose correlation is non-zero. I next estimate the second-step model and create the χ^2 difference test.

To claim that this χ^2-difference test is a clean and independent test of the structural part of the model is nonsense. This is just as unacceptable as looking at a correlation matrix in which all but one of the values are zero, and then deciding to use precisely that one location as one's hypothesized causal model. Clearly this type of use of the step-one model is unacceptable and can not be justified by the Steiger, Shapiro and Browne (1985) proof. Hence, it is abundantly clear that their proof being mute on the issue of data-prompted revisions does not justify a blanket assertion that the difference χ^2 test will always be acceptable, *no matter how one uses the first-step model.*

Our imagined researcher would be on more solid ground if he/she displayed a commitment to the initial structural model by refusing to change any part of the structural model on the basis of what was observed in the step-one measurement model. That is, if revisions are confined to the measurement portion of the model, then the problem is markedly reduced.

But a reduced problem is not necessarily a nonexistent problem. Can one do this in practice? Consider the advice provided by Anderson and Gerbing on how to proceed if one confronts a first-step measurement model that has unacceptable fit: "Relate the indicator to a different factor, delete the indicator from the model, relate the indicator to multiple factors, or use correlated measurement errors" (1988:417). Notably absent from this list, but equally reasonable, is an invitation to relate one or more of the indicators to a new factor (concept).

Now, if I displayed an absolutely dogmatic dedication to my original structural model, and made any of these five possible types of changes, I would have to describe myself as both theoretically vacuous and naive. There is some degree of theoretical ineptitude to doggedly sticking to an unalterable theory when one knows that more concepts are required, or that measures once thought of as pure indicators of one's concept are really mixtures of two or more concepts, or that what was initially taken as an indicator of one concept really should have been an indicator of quite a different concept. If one understands how the meaning of concepts change as the connections to the indicators change,[27] then any dogmatic dedication to an initially imagined theoretical structure is foolish once the measurement structure has been substantially altered. Hence, while an absolute dedication to the initially specified structural model might preserve the statistical independence of the χ^2-difference test, this purity would be purchased at a price of demonstrating one's rigidity, inflexibility, and probably scientific ineptitude![28]

We will gain a slightly different perspective on this when we discuss equivalent models in Chapter 4. The equivalence of various ways of addressing particular ill-fit covariances means that the inclusion of additional effects in the measurement model may render unnecessary some corresponding effects from the structural model. Hence, one can view the second-step structural model as becoming progressively more exploratory with each additional data-prompted change to the first-step model.

For the moment, let us tentatively side with Fornell and Yi, and conclude that the onus is now on Anderson and Gerbing to demonstrate why they believe that Steiger, Shapiro and Browne's (1985) failure to explicitly speak to the exploratory-confirmatory question should be taken as meaning that there can be no contamination of the χ^2 difference test, no matter what one does on the basis of the first-step measurement model.

Though this concludes our explicit use of Fornell and Yi's critique of the two-step process, we need to consider a few further points that do not surface in the 1992 exchange, before we make our final tally.

2.2 Extending the Debate

2.2.1 Nested Models

The Steiger, Shapiro and Browne (1985) proof is central to another, so far overlooked, challenge to the two-step procedure. The nested models addressed by Steiger, Shapiro and Browne (1985) are models in which placing constraints on one of the models produces the other (nested) model. The usual constraints are the replacing of a formerly free effect with a fixed zero or non-zero effect, or constraining two previously separately estimated coefficients to be equal. Neither the Steiger, Shapiro and Browne (1985) reference, nor any reference I know of, demonstrates that a model containing directed effects can be obtained by placing equality, or fixed (zero or non-zero) constraints on a correlational model.

We can highlight the difficulty with using constraints to convert correlational into directed effects models by imagining a directed-effects model which is structurally saturated, so that it contains as many coefficients as the correlational model from which it is supposed to have arisen. This model definitely is not nested within the correlational model. The degrees of freedom for the original and the supposedly nested models would be equal, and hence their difference is zero, but there is no such thing as a χ^2 with zero degrees of freedom! The point here is that if the models have equivalent numbers of estimated parameters, one cannot argue that constraints on the correlational model have done the translation from a correlational to a directed-effects model.

The difficulty is clear. Constraints are the essential component of any conceptualization of nesting, and hence are essential to the χ^2 difference test for nested models. What are the constraints that turn a correlational model (the free factor correlations at step one) into a directed-effects model (the directional structural effects at step two)? The zero degrees of freedom example above illustrates either that it is possible to introduce constraints without altering the degrees of freedom, or, if all versions of constraints consume degrees of freedom, that it is something other than constraints that does the translation from undirected to directed effects. Without a specification of the nature of the requisite constraints, the mathematical foundations to the χ^2 difference tests comparing first-step correlation and second-step directed effects models are resting on uncharted mathematical ground. The Steiger, Shapiro and Browne (1985) proof in particular cannot be relied upon because the styles of constraints it requires have not been demonstrated to be sufficient to translate a first-step correlational model into a nested directed-effects model.

A further awkward fact is that the supposedly nested directed-effects model contains variables not included in the original correlational model. Specifically, the error variables whose variances appear as diagonals in the ψ matrix have no counterparts in the correlational step-one model. The usual definition of nested models demands that the models are based on the same variables,[29] and hence the magical appearance of these new error variables within the supposedly nested directed-effects model also exceeds the usual definition of nesting. The Steiger, Shapiro and Browne (1985) proof of independence of a basic χ^2 and nested χ^2 difference is based on the usual definition, and hence again is not applicable to the move from a correlational to a directed-effects model implicit in the two-step procedure.

It may be possible to appropriately redefine nesting[30] to allow for the conversion of correlational into directional models, and for the emergence of error variables, and hence to obtain a proof of the independence of the original and difference χ^2 tests (if no model revisions whatever are prompted by the first step). But as yet, there is no source that demonstrates this, so the χ^2 difference test employed as the test of the second-step structural model is resting on an undefended leap of faith.

2.2.2 Measurement Modifications at Step Two

Let us return to the standard claim of the two-step procedure that measurement modifications (additional λ's and grudgingly θ covariances) should be made at step one while structural modifications are to be made at step two. Here we will approach this backwards by asking: *Why should modifications at step two be restricted to modifications that do not involve effects from the concepts to the indicators*? One might be inclined to think that there would be no need to make such modifications if the step-one model fit acceptably, but one would be wrong. Even if the step-one model fits, the step-two model may fail, and it is the failure at the second step that we are trying to overcome by considering what it is that keeps us from making further (second-step) modifications to the measurement (first-step) model.

Imagine a very sparse second-step model that is just a touch too sparse because it omits an effect from say η_1 to η_2. The failing second-step model can be saved by postulating some mechanism linking η_1 to η_2 that had not been anticipated. It is likely that this mechanism involves some conceptual variable that also had not been anticipated. How can proponents of the two-step procedure justify a claim that it would be acceptable to imagine a new concept as intervening between η_1 to η_2 (justifying the insertion of a new effect between η_1 and η_2) but unacceptable to entertain the possibility that the new concept also, or alternatively, intervenes between η_1 and η_2's indicators, and hence justifies new λ effects linking η_1 to η_2's indicators?[31] In both cases we are admitting to having overlooked a concept. Why should it be permissible to correct only one of the two kinds of overlookings at step two?

We should be clear that a fundamental component of this particular trouble is the sparseness of the second-step model. Factor analysis typically determines the *minimum number of concepts* required to account for the covariances among the items, *conditional on there being a fully free set of covariances among those concepts*. Our postulation of a markedly sparse model places substantial constraints on the covariances among the concepts and hence the minimum number of concepts found by the factor analysis may no longer be the true minimum number required to account for the item covariances. More concepts may be required if there are constraints on their covariances. The excess covariances among the concepts populating the first-step model may mask (obscure the diagnostics recommending) any need for more concepts in a theoretical world containing substantive constraints on the concepts' covariances (the sparse model).

The query we are posing is why the newly added concepts must never be allowed to influence any indicators? How could anyone determine, a priori, that insertion of another conceptual-level effect is acceptable but that allowing the new implicit intervening variable to influence any of the observed indicators directly is unacceptable? I do not see how this could be defended. The problem this poses for the proponents of the two-step procedure is that it again depicts the researcher as simultaneously confronting measurement and structural effects as a solution to model ill-fit. The proponents of the two-step process find this discomforting, while I view this as inevitable.

Note that the imagined sparseness of the second-step model is produced by using directed β and γ effects. Hence, if we grant the validity of the preceding claim, we are again siding with Fornell and Yi in claiming that it is structural effects and not merely the number and identity of concepts that are important for determining measurement structure (recall the discussion in Section 2.1.1).

2.2.3 The Meaning and Dimensionality of Concepts

2.2.3.1 The Approach Taken in *Essentials*

Most of the remaining issues in the two-step debate center on meaning. How can I tell you what my fictitious concepts mean? The essence of the answer provided in *Essentials*, and elaborated in Section 1.5, is that I can tell you most clearly what I imagine an abstract concept to mean by *telling you which of the observed items (indicators) are most similar to my conceptualization*. My a priori commitment to some items as closer to, more similar to, or more relevant for my conceptualization, most clearly links my concepts into your world because you see precisely which parts of our shared observable world I attach my conceptualization to. It is when researchers attempt to avoid making such commitments that the problems mount. Saying that "what I mean by a concept is something, anything, that is common to this set of items" (the standard justification for factor analysis) raises the issue of whether what you see as common is necessarily what I see as common. There may be more than one characteristic common to the items, hence we enter the dimensionality debate to assure ourselves that we might not be routinely pointing to two different characteristics or dimensions as underlying the items.

If the "something common" is sufficiently clearly presented to allow any reader to create a nearly infinite set of equivalently good items that would be both agreeable to the researcher and the reader, then this approach creates little problem. Unfortunately "investigators do not typically begin with a clear theory of content that enables them to define a (virtually) infinite set of tests which only time prevents them from creating and using" (McDonald, 1981:108). Given the diverse substantive areas to which structural models are being applied, and the limited theoretical-diligence of publication oriented number crunchers, it is unlikely that mere admonitions will alter this aspect of current operating practice.

2.2.3.2 Meaning in Two Steps?

The issue of the meaning of concepts returns us to the basic claim that using a first-step correlational model keeps one's theoretical hands off the data (Section 2.1.1). Let us try this again, but this time start from Anderson and Gerbing (1988):

> In the presence of mis-specification, the usual situation in practice, a one-step approach in which the measurement and structural submodels are estimated simultaneously will suffer from *interpretational confounding* (c.f. Burt, 1973, 1976). Interpretational confounding "occurs as the assignment of empirical meaning to an unobserved variable which is other than the meaning assigned to it by an individual a priori to estimating unknown parameters" (Burt, 1976, p. 4). Furthermore, this empirically defined meaning may change considerably, depending on the specification of free and constrained parameters for the structural submodel. Interpretational confounding is reflected by marked changes in the estimates of the pattern coefficients when alternate structural models are estimated. (418)

I agree that the meaning of concepts change as the λ's (pattern coefficients) change. I half agree that interpretational confounding (alterations in λ's) is likely to accompany changes in one's structural theory. This is true only *if one has not adopted the fixed 1.0 λ and θ procedure recommended in* Essentials *and extended in Section 1.5*. Scaling concepts with a fixed 1.0 λ and corresponding fixed θ variance allows no, let alone marked, changes in the coefficients linking the concept to its scaling indicator. These fixed coefficients will not change "depending on the specification of free and constrained parameters for the structural submodel." The stability of this link between the concept and the scaling indicator provides a solidity to the meaning of the concept that is unattainable by any strategy that allows all the λ's for a concept to be free. The researcher begins by forcing each concept to take on the meaning he or she intends (presumably the meaning most consistent with the researcher's current theoretical views), and hence the chances of interpretational confounding are eliminated, or at least minimized, before any estimation is attempted.[32]

Let us play along, however, and temporarily assume that the researcher has not used the fixed λ and θ procedure, and hence that the statements ending the quote pertain: namely that the estimated λ measurement coefficients, and hence the meaning of the concepts, would be conditional on the structural model used. This seems to argue that researchers can avoid interpretational confounding by becoming committed to a single structural model since only one meaning is then being used. But Anderson and Gerbing continued by making exactly the opposite claim:

> The potential for interpretational confounding is minimized by prior separate estimation of the measurement model because no constraints are placed on the structural parameters that relate the estimated constructs to one another. (1988:418)

I view this as being exactly backwards. To me, the more constraints we place on the concepts, the clearer they become because the more rigidly and tightly they are held within the confines of the other concepts. I know most clearly what a researcher means by a concept when it is embodied in a predetermined set of structural coefficients (β's and γ's), not when that concept is allowed to float freely between being a cause or an effect, as is possible when a merely correlational structure links the concepts. We know concepts more clearly when we know how they fit in *a particular* cause-effect system, than when they are involved in *some as yet unspecified* causal system (namely any one of the multitude of causal systems giving rise to the correlations among the concepts in Anderson and Gerbing's first step). Clearer meanings on concepts arise when the concepts are imbedded in a particular theory which the researcher operationalizes as β and γ coefficients.

We now see two specific ways that *the meanings of concepts are clearer if one adopts the one-step approach. The fixed 1.0 λ's and θ variances provide a solid connection to the observed variables that most closely approximate the researcher's theoretical conceptualizations, and the structural model forces these concepts to be placed into a clear and unambiguous context, thereby allowing minimal conceptual ambiguity*. We have the best of both worlds: a clear statement of how the researcher views his or her concepts as connecting to observable events, and a clear statement of how the different concepts connect to one another. What more could one possibly want in attempting to determine what the researcher means by a concept?

Contrast this with the first step of the two-step approach, with its lack of commitment to a particular theory implicit in the use of merely correlated concepts, and its weak commitment made in seeking something, anything, common to each set of indicators. Yet it is this first step that is supposed to determine the measurement of the concepts! Are we really to believe that a strategy with such a loose grasp on the concepts' meanings really lives up to its advance billing of determining the measurement structure? I do not see how this is possible. I feel much more confident that I know a researcher's meaning on a concept when the researcher tells me which indicator most closely approximates the concept's meaning, and how this particular concept fits into an overall conceptual structure.

For another perspective on meaning in the two-step procedure, consider again the options Anderson and Gerbing (1988:417) see as being available to researchers whose first-step measurement model fails. One may "relate the indicator to a different factor, delete the indicator from the model, relate the indicator to multiple factors, or use correlated measurement errors." Anderson and Gerbing believe that either of the first two options are preferable because the third and fourth reduce the potential for unidimensional measurement and hence "obfuscate...the meaning of the estimated underlying constructs."

We agree that adopting the third and fourth options alter the meanings of the constructs, but so do the first and second options.[33] Option one, relating an indicator to a different factor, should alter the meaning of the concept from which it is extracted by reducing the domain of meaning granted to that concept. Some specific portion of the domain of applicability that the researcher had been granting to the concept has been found to be unacceptable, and hence the meaning of the new concept covering the remaining items must acknowledge this reduction in domain of meaning. The domain of meaning of the concept to which the indicator has been added must correspondingly expand to acknowledge the new domain segment that is now required to cover the imported item.

Deleting an indicator from the model, option two, similarly demands a reduction in the domain of the concept losing the indicator. The new meaning specifically must not include the deleted item. But the item is no longer in the analysis to keep the researcher honest. The item has been dropped, so there is nothing to keep the researcher from using the same old meaning as before. The researcher's original, now inappropriate and obfuscating, meaning for the concept is likely to persist.

Clearly, all the options for changing a measurement (step-one) model afford possibilities for obfuscating the meanings of the concepts. But note that if we are to follow the Anderson and Gerbing two-step approach, the changes in meanings are specifically not to be informed by any theoretical (structural effect) considerations, and these changes should not alter the theory to be tested. The concept with the altered meaning is supposed to display the same causal response or causal effectiveness as it was initially postulated to display. Again, the essence of the two-step approach is that the measurement and structural concerns can and should be addressed in isolation. So we are to continue imagining the same old effects as linking the changed concepts! Are our theories really so imprecise that they can work with any old definition on the concepts!

We end by paraphrasing the quote from Anderson and Gerbing (1988) with which we began this section.

> In the presence of mis-specification, the usual situation in practice, a *two*-step approach in which the measurement and structural submodels are estimated *separately* will suffer from interpretational confounding. Interpretational confounding occurs as the assignment of empirical meaning to an

> unobserved variable which is *other than the meaning assigned to it by an individual prior to estimating unknown parameters*. (text in italics added or altered)

2.2.3.3 Unidimensionality and Meaning

We pause to consider the dimensionality of items because it provides a glimpse into the mental framework within which Anderson and Gerbing were functioning when they attempted to revive the two-step procedure.[34] Let us begin with a quote that appears in their "Confirmatory Measurement Models" section, prior to their defense of the two-step debate. They state:

> A necessary condition for assigning meaning to estimated constructs is that the measures that are posited as alternate indicators of each construct must be acceptably unidimensional. That is, each set of alternate indicators has only one underlying trait or construct in common (... McDonald, 1981). (1988:414)

Roderick McDonald (1981:113) defines a *set of items* as being unidimensional "if and only if the set fits a (generally non-linear)[35] common factor model with just one common factor." If two factors are required, the item set is said to be two dimensional, and so on. Note that this definition describes the dimensionality of a set of items as a whole, not the dimensionality of a particular item, and that the definition itself is mute about any research preference for unidimensional, over bi- or multidimensional items or sets of items. So Anderson and Gerbing's preference for unidimensional sets must be predicated on some other implicit argumentation.

Let us try to get a grasp of their position by considering another definition of dimensionality that shifts the emphasis away from the factor model (Anderson and Gerbing's context) toward the regression model. Let me define the dimensionality of an indicator item as the number of non-error causes of that item.[36]

There is a sense in which every indicator variable is unidimensional. The values of a variable reflect higher or lower values on *a* dimension, where the unitariness of the dimension is warranted by the unique methodology producing the set of values assigned that particular name. My definition takes this as given, and treats the issue of dimensionality as a query about the number of sources of, or determinants of, a particular indicator. It is a question about a model (as is McDonald's), and it asks us to count the number of concepts having non-zero λ's in the row of Λ corresponding to the equation for the indicator of interest.[37]

Consider now a shared indicator, that is, an indicator loading on two underlying concepts where each of those concepts is clearly meaningful because it has been well identified with a fixed 1.0 λ and small θ as discussed above. (You might imagine one of the concepts as the gender of the respondent, the other as his or her elusive "attitude toward *X*.") The equation for the indicator sharing these two conceptual sources is like a multiple regression equation where the concepts act as predictor variables and the indicator is the dependent variable. Is there anything particularly mysterious or difficult to understand when a dependent variable is dependent on two or more predictor variables? From this perspective neither the concepts nor the slopes (λ's) seem particularly disconcerting or difficult to understand. The λ linking the attitude indicator to the gender concept is some form of gender effect, in that one gender provides higher indicator scores, while the other λ reports on the degree to which "attitude *X*" makes itself apparent in the indicator. This should be straightforward for anyone with a background in multiple regression.

Why then is it that the "*necessary condition* for assigning meaning to estimated constructs is that the measures that are posited as alternate indicators of each construct must be *unidimensional*" (Anderson and Gerbing, 1988:414, emphasis added)? Defining the dimensionality of an item as the number of predictors in the equation for that indicator, in a direct parallel to multiple regression, lets us agree that "measurement models that ... have indicators that load on more than one estimated construct do not represent unidimensional construct measurement" (1988:415) but disagree that this is necessarily "problematic" (1988:415) or that it obfuscates "the meaning of the estimated underlying constructs" (1988:417).

It would be simpler for the scientist if each observed variable tapped a single conceptual source, but it is the real world, not the researcher, who determines whether any particular measure is or is not unicaused. The demand that sets of unicaused variables be located for each concept in a model prior to proceeding is not only a formidable task, it is an unnecessary task. To escape this particular illusion we had to switch out of factor mode thinking and into regression mode, and to use the fixed λ and θ procedure. Neither of these is particularly difficult but together they are sufficient to dispel the sense of necessity underpinning the above quotes. So yet again we catch the factor model deluding us about what "must" be done.

2.2.3.4 Concepts with Causal Indicators or No Indicators

Proponents of the two-step procedure have difficulty with models containing concepts measured with causal indicators, or concepts having no direct indictors. If a model incorporates observed variables that indicate a latent concept by causing that concept (Bollen and Lennox, 1991; MacCallum and Browne, 1993), then a first-step factor model is inappropriate because it is misspecified. The causal sequence required by the researcher's model is that the indicators cause the concept, and biased and entirely invalid estimates will result if the researcher adopts the now-inappropriate sequencing that the concept causes the indicators implicit in the first-step factor model.

Models containing concepts with no direct indicators (Hayduk, 1990; Chowdhury, 1991:200; Germain, 1994:52; Thon, 1993; and see Chapter 3 below) are just as antagonistic toward the first-step factor model. These models contain concepts whose meanings are entirely derived through their causal connections (β and γ effects). Given that the first-step factor model cannot accommodate correlations with a concept having no indicators, this is another style of model for which the two-step approach is simply inappropriate.

The inappropriateness of the two-step approach for models containing indicatorless concepts does not make such models wrong or uninteresting. Chowdhury's (1991) model, for example, concerned the effects of three precursor demographic factors (exposure to intercourse, breastfeeding, and infant and child mortality) on couples' deliberate control of fertility. The substantive demographic issue was whether the three causes were acting via separate causal mechanisms or whether they could be operating via a common mechanism which the demographers had been calling the potential supply of children. Chowdhury (1991) first estimated a baseline model with disparate direct effects from the precursors (the dashed lines in Figure 2.5A) and then reestimated the model with this segment of the model respecified to include an indicatorless carrier mechanism (the η_4 concept, potential supply of children) as required by demographic theory.[38] The resulting estimates provided an equivalent model to the baseline model (essentially the same fit and diagnostics as the base model) and demonstrated that the data could reasonably be provided an interpretation involving a phantom indicatorless η_4. The estimates also provided the effect sizes required for consistency with this particular alternative interpretation.

Figure 2.5B provides an example of another style of indicatorless model likely to be of interest to political scientists studying countries like Mexico where fraud may contribute a substantial component of official vote counts. We can imagine state or precinct official vote counts as including both true and fraudulent votes. The errorless sum of the true and fraudulent votes must equal

Figure 2.5 Indicatorless Concepts.

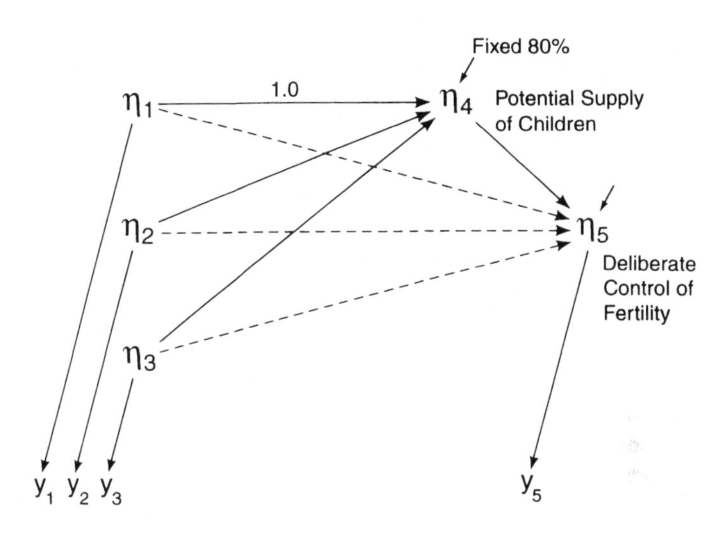

A: A demographic example.

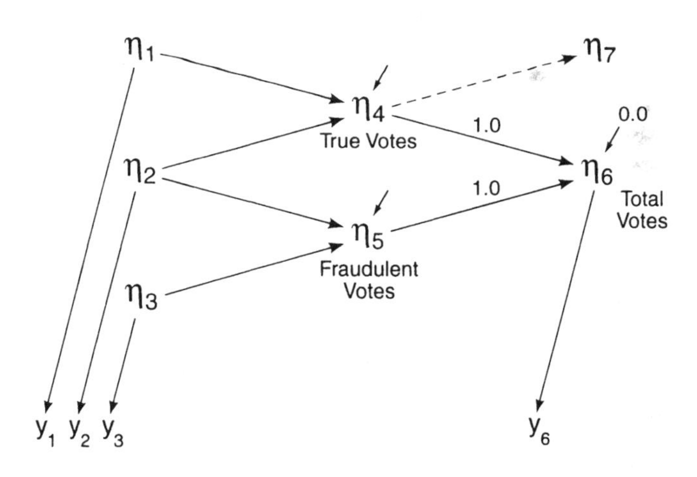

B: A political science example.

the total count (there are no other types of votes), so the model includes two fixed 1.0 values (each true and fraudulent vote contributes exactly 1.0 to the total count) and a fixed 0.0 error variance (all votes are either true or fraudulent).[39]

As it stands, the model segment in Figure 2.5B is underidentified but the styles of information that would assist identification are clear.[40] In the absence of sufficient additional constraints to identify the model, one might have to resort to a sensitivity analysis, but even this would focus attention on questions that would at best be implicit in a model failing to distinguish between true and fraudulent mechanisms of influencing vote tallies in seeming-democracies.

If you judge this style of model to be novel, original, creative, or at least a prod to the sluggish imagination, you should view Anderson and Gerbing's suggested two-step approach as stifling because their procedure would preempt the consideration of such models.

2.3 So What?

2.3.1 The Two-Step Approach in Review

The preceding comments suggest a downgrading of the two-step approach from being so obviously acceptable it should be routinely encouraged, to a procedure that is sometimes acceptable. Proponents of the two-step procedure certainly have not been clear about what they demand at the first and second steps. And I suspect the clearer they become in terms of prescriptions, the more opaque the advantages of the two-step procedure will seem to practicing researchers. Some specific problems with the procedure and limitations on its use are:

1. The two-step procedure is inappropriate if even one concept has fewer than three indicators (Section 2.1.1.2).

2. Confirmatory, not separate, factor analyses must be used at the first-step (Sections 2.1.2.2 and 2.1.3.1), and a fixed 1.0 λ and a corresponding fixed θ should be used for each concept at both steps one and two, to demonstrate that specific and consistent meanings are assigned to your concepts at both steps (Sections 2.2.3.1 and 2.2.3.2).

3. Scales based on factor analyses are doubly problematic. Using a scale to measure a concept violates the requirement of at least three indicators (point 1 above) and both the scale and the indicator items themselves must be modeled if one is to respect the causal forces underlying the scale's construction

(Section 2.1.3.1.1). This expands the size of the model and may lead to colinearity problems.[41]

4. The two-step procedure specifically precludes any possibility of allowing a sparse theory to assist in identifying a dense measurement structure (Section 2.1.3.2). The procedure can be used only if the step-one model is identified and has a χ^2 with at least 1 degree of freedom; otherwise there is no first-χ^2 to use in creating a difference-χ^2. And the structural model must not contain as many coefficients to estimate as did the step-one measurement model; otherwise the difference-χ^2 ends up with 0 degrees of freedom.

5. Do not even consider the two-step procedure if you have a model containing a concept that has no indicators (Section 2.2.3.4).

To this list we might add some statistical cautions, which seem so straight forward that we have not addressed them separately above.

6. Stop, and never go on to step two, if the first-step model fails. ***For all models, whether part of the two-step procedure or not,*** *we recommend using a χ^2 $p>.75$ or so (instead of .05) to acknowledge our implicit favoring of the null hypothesis.*[42] If the proponents of the two-step procedure were clearer about this we might not find researchers with impressively significant step-one χ^2's proceeding to step two.

7. No step-one modification indices should exceed 4.0[43] as this indicates that specific significant measurement modifications are required, and you would be behaving inconsistently if you did not solve all your measurement problems at step one before going on to step two. With multiple indicators, the modification indices for the fixed θ variances are pivotal because freeing these demands corresponding changes in conceptual meanings.

8. Use no free measurement parameters (λ's and θ's) at step two. Changes in the estimates of these values are changes in concepts' meanings, and you are not allowed to change the meanings of your concepts between steps one and two. Without this, there is no substance to Anderson and Gerbing's view that "the measurement model should change only trivially, if at all, when the measurement submodel and alternate structural submodels are simultaneously estimated" (1988:418). You will have to reduce your degrees of freedom for your second-step test by one for each measurement parameter fixed on the basis of the first-step model's peek at the data. Hence, you will either have to look up the significance of your second-step model χ^2 instead of depending on LISREL's reported χ^2 probability, or run LISREL 7 or 8 where you can reduce the degrees of freedom by specifying *DF=-#* on the output line.

And we might even add a few admonitions to the list.

9. Report all model changes made at both step one and step two so the reader can assess, and hence mentally adjust for, the nature and extent of exploratory compromising.

10. Muster the courage to risk being placed in a damned if you do or

don't situation. The second-step χ^2-difference becomes indefensible if the structural model is altered on the basis of step-one information; hence, if you change the measurement model, you will be criticized no matter what you do at step two. You will either be criticized for displaying conceptual rigor mortis (applying the same old effects to a whole new set of concepts, Sections 2.1.4 and 2.2.3), or for conceptual waffling (failing to stick to your original structural claims). If you really were just exploring, you will have considerable difficulty justifying why your explorations focused first and preferentially on the λ's for some irrelevant correlational model, rather than giving uniform and unbiased attention to all the coefficients (measurement and structural) in the model containing your best guesses about a reasonable conceptual structure (your initial β's and γ's). So cross your appendages and hope no measurement changes are required at step one.

11. Pray that some statistician gets around to modifying Steiger, Shapiro and Browne (1985) so that the χ^2-difference test for the move from a merely correlational model to a directed effects model rests on something other than blind faith.

2.3.2 Recalling this Chapter's More Fundamental Themes

While the above characterizes this chapter in terms of its specific relevance for the two-step debate, focusing specifically on the two-step debate misses much of what I view as important in this chapter. A few general principles ground many of the critiques and comments.

A primary theme has been the controlling of concepts' meanings by using one fixed 1.0 λ and fixed corresponding θ for each concept. The fixed λ procedure for scaling each concept is now generally accepted. The fixing of the corresponding θ or error variance is not yet widely adopted. This probably reflects both a lack of appreciation of how this clearly and rigidly assigns meanings to the underlying concept,[44] and a sense of discomfort at specifying measurement quality as opposed to assessing or estimating measurement quality. This sense of unease fails to recognize that it is multiple indicators of the concepts, the indicators with free λ's and θ's, that preserve the possibilities for assessing and testing the measurement structure imposed by a fixed λ and θ. With a single indicator, the best the researcher can do is assert a clear meaning for each concept. With two or more indicators, the proportionality constraints discussed in Chapter 1 are rigid and unforgiving constraints that permit one to assess one's assertions about the connections between one's conceptual meanings and the available indicators.

In fact, being abundantly clear about the meaning one attaches to a concept by fixing both a λ and the corresponding θ for each concept can provide a more stringent test of the postulated measurement structure. Imagine a concept with four indicators. If we fix only one λ at 1.0 we are left with a total of seven coefficients (three λ's and four θ's) to account for this portion of the measurement model. If we fix one λ at 1.0 and fix the corresponding θ at some specific value, there are only six free coefficients to account for the same portion of the measurement model. The extra inflexibility, rigidity or precision in specifying the meaning of the concept is paid back as an additional degree of freedom and as a tighter, more parsimonious, story about the functioning of this set of indicators. While one could estimate all four θ's, one pays for it. One pays the price of a degree of freedom, and one pays the price of reporting to the world that one's conceptualization bore no specific connection to even one indicator. One may even pay the price of having to live with an estimate that implies a conceptual meaning one finds objectionable.

It should be obvious that estimating all the θ's for a concept weakens the researcher's commitment to the model, and hence contributes to the current milieu of models dropping like flies, to no one's real concern. As many, or more, models will fail if fixed θ's are used. The difference is that the level of concern about failures should rise and, hopefully, prompt the substantive reconceptualizations that could reduce the discipline-wide model failure rate in the long run.

Another theme has been the inseparability of measurement and theory. Our concepts are granted meaning both through their links to the indicators and by their links to other concepts; hence to attempt either of these separately is attempting to defend partially determined meanings. I am willing to accept correlations among concepts as theory, but I view β and γ effects as being much stronger theory because effects are capable of implying correlations. Specific effects imply specific correlations,[45] but the reverse is not true. Specific correlations by themselves do not imply specific effects. Hence, if I have a choice, I will theorize using effects rather than correlations. And when I am asked to theoretically specify the meaning of my concepts, I will do so with the strong effect version of theorizing whenever possible, and only as a last resort invoke a weak correlational theory.

For a slightly different perspective on this, consider whether you agree with the assertion that we only know what we are measuring when our concepts, and hence measurements, behave in relation to one another as they theoretically should. Now, ask yourself if you would be willing to discard a step-one measurement model with an acceptable χ^2 if it contained a large correlation, or a null correlation, between two concepts that you would have difficulty accounting for, where that accounting was specifically never to include any reference to the effectiveness of any of the concepts (i.e., no β or γ effects). That is, would you be willing to use the estimated ϕ values to reject a step-one measurement model having an acceptable χ^2? If you believe that the estimated ϕ coefficients would be insufficient justification for discarding a step-one measurement model that fits the data, you are saying that the covariances between the concepts are too weak theoretically to ascertain how the concepts should behave, and hence you are abandoning the claim that heads this paragraph if you use step one of the two-step procedure. If specific correlations among concepts are too analytically weak to justify rejecting a model, correlations are too weak to provide the conceptual interconnectedness necessary for grounding confident assertions about measurement.

A final theme has been that the researcher has no less a stake in the measurement submodel than in the structural submodel, and that it is no easier to change one part of the model than the other. The researcher is clearly backing off his or her conceptualization of the substantive domain as soon as either form of model modification is made. The researcher is free to modify either the measurement or structural submodels to accommodate any particular ill fit covariance but both types of changes provide substantial challenges to the researcher's theory. To change the connection between a concept and an indicator, or to change the connection between two concepts, similarly invokes the need for another mechanism. If anything, it is more, not less, devastating if that additional mechanism leads directly, rather than merely indirectly, to other observed variables. In either case, the meanings of the concept are only minimally changed if the fixed λ's and θ's are in place. It is the existence of a previously unrecognized mechanism that is theoretically threatening.

The combined measurement-structural modeling approach prods us to question whether measurement structure must, or even can, be determined prior to consideration of the structural model. This is frequently asserted (Anderson and Gerbing, 1982, 1988, 1992; Hunter and Gerbing, 1982), and indeed, it is fundamental dogma in the factor analytic tradition. But merely repeatedly asserting this is no longer adequate. The implicit challenge in the preceding pages is that this must be defended or it will fail, and if it fails, so fails several decades of psychological literature. If we back off confirmatory factor analysis as an acceptable first step, then the many years of psychological scale construction via factor analysis are first steps to nowhere!

I view the two-step debate as a temporary distraction to be endured but ultimately disregarded as our attention turns to more useful pursuits. Much more is to be gained by encouraging careful, precise, and attentive modeling habits, especially if this is spiced with a substantial dose of that illusive quality called creativity.

Notes

1. More recent proponents of the two-step approach are Lance, Cornwell and Mulaik (1988), Cohen et al. (1990) and Scheines et al. (1994). Kumar and Dillon are among the few who have argued that "though measurement and structure *can* be

evaluated independently of each other, in general, they *should not* be" (1987:98, emphasis in original).

2. Some of the original conceptual covariances may remain as ϕ covariances.

3. Recall *Essentials* (112-116).

4. A sparse structural model can disproportionately add degrees of freedom to the χ^2 test and increase its probability but the size of the residuals should not decline.

5. Concepts with single indicators, or no direct indicators at all, are possible. See Section 1.5, Section 2.2.3.4, *Essentials* (213), and Hayduk (1990).

6. I treat as minor Fornell and Yi's repeat attack on the claim that the first step is theory-free and Anderson and Gerbing's repeat rejection of this. The set-specific nature of measurement error is not really minor, but fits more appropriately into the debate about the meaning of concepts which we examine in Section 2.2.4.

7. A saturated model is one in which as many coefficients have been estimated as there are data points (covariances) to be accounted for. Here we are referring to the saturation of only the conceptual level of the model, where the variances and covariances among the concepts are to be accounted for and the corresponding free variance and covariance coefficients provide a perfect fit by a one-to-one correspondence. Model B has as many structural coefficients and necessarily provides an equally perfect fit to the concepts' covariances. No indicators for these concepts are shown, as the indicators are not at issue.

8. Many more equivalent models can be created by turning the covariance between η_1 and η_2 in model B into a directed effect, and then again randomly allocating the η variables to the various possible positions within this fully-directional saturated model.

9. Fornell and Yi (1992a) question Anderson and Gerbing's (1992) assertion that model C and model A are also equivalent. After you have read the discussion of equivalence in Chapter 4, you should be able to convince yourself that Anderson and Gerbing are indeed in error on this point, but this is irrelevant to the basic issue, so we let this pass.

10. This can be hand-calculated from Equation 4-62 in *Essentials*, or by entering the fully fixed model into LISREL and examining the Σ or moment matrix.

11. The other separate factor model with its mistaken indicator is similar to our model 1.2D, but again the similarity of the proportionality constraints renders the mistake undetectable as long as one is building a model with only a single common factor and examining only the covariances among the four indicators. Note also that the "intervening variables" do not function as common causes of more than one indicator, so the necessarily proximal nature of common causal factors (Section 1.4) is not at issue.

12. This parallels our model 1.2E, with the concepts either being correlated or causally linked. Switching indicators between the concepts renders either of these models incapable of matching the double-proportionality pattern characterizing the true behavior of the indicators. Both the causal and correlational model imply a particular type of double-proportionality among the indicator covariances; it is just that the proportionalities implied by the models will not match the proportionalities observed in the data because they are misplaced, or shifted away from their true location (a row in the between-factor portion of the item covariance matrix becomes a column when an indicator is mistakenly included under the wrong concept).

13. **Unbiased** means that the average of the estimates calculated for repeated random samples from a population (i.e., the average of the sample statistics) equals the numerical magnitude of the characteristic of interest in the population (the population parameter). This guarantees that though any particular sample's estimate may be above or below the true population value, on average there is no bias towards over- or

underestimating the magnitude of the quantity of interest in the population.

Consistent means that as sample size increases, the sampling distribution of the estimator progressively squashes in around the true population parameter's value. The variance of the sampling distribution progressively decreases and the sampling distribution is "on target" in that it becomes increasingly certain that the estimated value is within any small arbitrary distance from the value of the population parameter (cf. Bollen, 1989a:467). This guarantees that as sample size increases, more precise estimates of the population parameter can be made (the length of the interval in "interval estimates" declines) and that more precise hypothesis tests can be made (the length of the interval defining the acceptance region around the null hypothesis declines).

Efficient means that, of the different estimators (formulas) that might be used in an attempt to predict the population parameter of interest, one estimator (the most efficient estimator) has a sampling distribution with smaller variance than the variance of the sampling distributions for any of the other estimators (for every specific sample size). Using the most efficient estimator guarantees that one is making the most precise statements that one can (using the smallest confidence interval or null hypothesis acceptance region) and hence that one is not wasting research dollars by wasting the discriminatory power provided by the sample one has so laboriously amassed.

14. Fornell and Yi also argue that by ignoring the covariances between sets of indicators (rather than among each set individually), one is ignoring useful information and reducing efficiency. Since we have discussed this in conjunction with Assertion 2, we do not repeat this here.

15. This is a special case of the bias introduced whenever one ignores correlated yet effective predictors in multiple regression. See *Essentials* (46-48) for a discussion of why regression slopes change when correlated and effective predictors are added to regression equations.

16. Researchers knowingly avoiding inclusion of multiple indicators because this increases the likelihood of their model failing deserve to be convicted of professional misconduct in the court of academia.

17. The details of how Figure 2.4B is a simplification of Figure 2.4C run as follows. The equations for how each of the items arise in the Concept-X and a unique error (Figure 2.4C) can be inserted into the corresponding locations for the items in Eq. 2.7. That is, the items as determined by the real world are forced to enter into the scale the researcher has causally constructed. This provides

$$\text{Scale-}X = 1.0(\lambda_1\text{Concept-}X + e_1) + 0.5(\lambda_2\text{Concept-}X + e_2) + 2.0(\lambda_3\text{Concept-}X + e_3),$$

which can be rearranged to

$$\text{Scale-}X = (1.0\lambda_1 + 0.5\lambda_2 + 2.0\lambda_3)\text{Concept-}X + (1.0e_1 + 0.5e_2 + 2.0e_3).$$

Relabeling the sum of the indirect effects of the Concept-X on the indicators as the total effect λ, and the sum of the error contributions as e provides

$$\text{Scale-}X = \lambda\text{Concept-}X + e,$$

which is the equation for Figure 2.4B.

From this it is clear that the error e on Scale-X in Figure 2.4B originates from the errors on each of the three indicators in Figure 2.4C. Hence, it is imprecise to say that e in Figure 2.4B represents other (non-item) variables which cause Scale-X. The e

represents other variables that cause the items, and it is only through the action of the items (the 1.0, 0.5 and 2.0 weightings) that these error effects indirectly arrive at the Scale-X. This provides yet another reason why one should not interpret a concept to scale-indicator relationship without explicit discussion of the items comprising the scale.

18. The separate factor model might fail because one indicator affects another or because two items share a cause other than the Concept-X, but we need not be concerned with these reasons for failure since we are assuming that the separate factor model "justifying" the scale construction has worked.

19. The prime diagnostics would be the modification index for a λ if the background variable was a model η, or a residual if the background concept was exogenous while the Scale-X was attached to an endogenous concept.

20. Hence one correspondingly finds smaller T values and weaker statistical significance for the estimated coefficients (*Essentials*, 173-175).

21. In the 1992 exchange, the participants are actually focusing on a specific model at this point in their debate. I think Anderson and Gerbing (1992:328-329) err when they argue that all the six covariances in the first step of this particular model are relevant and hence not superfluous because testing would demonstrate that each is individually significantly different from zero. The covariances will indeed be significant, but this renders the coefficients relevant only from the perspective of the first-step model. Fornell and Yi were focusing on a second-step model that truly requires only four structural coefficients to account for all six non-zero covariances among the concept. This misperception takes Anderson and Gerbing off target at this juncture, so we ignore the details of this part of the exchange.

22. For examples, see Offodile, Ugwu and Hayduk (1993), or the interaction and nonlinearity models in *Essentials*, Chapter 7.

23. In their critique of Assertion 2, Fornell and Yi (1992a:304-306) provide a model where two additional degrees of freedom gained by conceptual sparseness are just sufficient to change a formal statistical decision about the model from reject to accept. This is a different, and weaker, style of challenge than I am proposing here. Fornell and Yi's style of objection can easily be overcome by adopting a flexible approach to statistics which avoids rigid decisions on the basis of test statistics that are just barely inside or outside the acceptance region. All one need do is report the α probability for the test statistic and *use it as a probability* in one's own thinking, instead of anointing this with any false certainty by having made a decision.

Models in which constraints among the conceptual level variables are required to make the model identified also provide a counter-example to Anderson and Gerbing's (1992:322) claim that "when the initial measurement and structural submodels are each correctly specified, the one- and two-step approaches are the same in the inferences that they would provide." If one model is underidentified and the other is not, they certainly would not be expected to provide the same results!

24. As explained in *Essentials* (163-167).

25. The rebuttal Fornell and Yi (1992b:337) provide to Anderson and Gerbing's (1992) critique of their challenge chides Anderson and Narus (1984) for interpreting each of two nested but basic χ^2 tests (i.e., not one basic and one χ^2-difference test) as independent tests (which clearly is unacceptable). This, however, is not the most fundamental aspect of the original challenge posed by Fornell and Yi, because they clearly state that "chi-square *difference* tests would be appropriate when all models in the sequence are hypothesized a priori, but *not* when some of the models are developed after looking at the data" (1992a:314 emphasis added).

26. Recall *Essentials* (167).

27. See Burt (1973, 1976) or *Essentials* (120).

28. If the theory at issue is someone else's, then one at least can justify one's rigidity on the basis that one is being true to that other person's theory. Here there may be some possibility of retaining a pristine structural model, even in light of measurement model revisions. One will then necessarily seem less than persuasive as to why one's initial measurement specification should have been viewed as acceptable from the perspective of the theory.

29. Actually, it demands that the nested model's parameters be a subset of the original model's parameter. So to be technically precise I should word this as stating that there were no ψ parameters in the original model, and hence no such parameters allowed in the nested model.

30. For example, by specifically addressing models whose parameters are composites of the parameters in the nested model, or by simultaneously addressing the issue of model equivalence and nesting. Another way to attempt to salvage the independence of the χ^2 test would be to claim that the first-step model being tested really is the saturated directed-effects model containing all the directed effects from the second-step model plus randomly assigned directed effects in the remaining locations required to achieve saturation and hence equivalence to the correlational model. This confronts the problem of how one selects *this* rather than any other one of the equivalent models. This selection cannot be justified or motivated without reference to the second-step model. This in turn demands a consideration of directional effects at step one, which again is precisely what the proponents of the two-step procedure are trying to avoid. Hence this would be awkward for the two-step proponents, even if it proved to be an acceptable strategy.

31. The new concept may also have effects on other model concepts that bypass η_2, but this is tangential to the current point.

32. It goes without saying that I disagree with Anderson and Gerbing's claim that a preferred way to set the metric of factors is to "fix the diagonal of the phi matrix at 1.0, giving all the factors unit variances" (1988:415). Until recently this approach could not be applied to all the concepts in a model because some other strategy was required to scale endogenous concepts. The η variables' variances are not model coefficients that can be set, but are instead implied by the estimated values of the effects, variances and covariances of the causal variables and error variables (*Essentials*, 111-112). Though one can now do this (Joreskog and Sorbom, 1993), I advise against it, precisely because it is accompanied by a set of entirely free λ's which permit unwarranted changes in the meanings of the concepts.

33. A fifth option of adding a concept also demands changes in the conceptual meanings of the previously existing concepts because their domains of applicability are now more constrained.

34. The perspective is also evident in Hunter and Gerbing (1982).

35. McDonald's next comment down-plays the nonlinear possibilities by claiming that "a simple, appropriate, and ancillary assumption is that the regressions of the tests on the factor are linear" (1981:113).

36. This definition implies that both the random and omitted unique variable interpretations of error variables do not contribute to the dimensionality of an item. But a unique cause of an observed variable that was entered as a non-error predictor of that variable would be counted in determining the dimensionality for that item, even if that cause was truly unique in that it received effects from, or sent effects to, no other variables in the model. If a researcher is sufficiently committed to a dimension to name it and acknowledge it as part of his or her theory, it becomes a countable dimension. I can imagine a unique cause of an observed indicator being interpretable (even if it has no other indicator) if, for example, its effect is involved in an equality or proportionality constraint with some other model coefficient. The unique cause could act independently of, but with a force equivalent to, the force of some other modeled variable.

37. One way to distinguish between McDonald's definition and my definition is to note that my definition is applicable even if there is a single indicator per concept, while McDonald's definition is not applicable in this instance.

38. The scale of η_4 was set by referencing it to the strength of one of the strongest suspected causes and the error variance for η_4 was examined using a sensitivity analysis (*Essentials*, 148) and ultimately fixed at about eighty percent of η_4's variance.

39. In general the error variables attached to unmeasured concepts and causally indicated concepts will be both non-zero and underidentified. Contrary to MacCallum and Browne, who propose that the residual variance should "be fixed at zero" (1993:540) to gain identification, I would propose first fixing the error variance at the variance the researcher would anticipate. This is more realistic and it would also lead to identification. It would also prompt one to consider supplementing this with a sensitivity analysis in which higher and lower fixed values could be entered to determine the sensitivity of the other model estimates to this particular component of the model's specification.

40. Causal variables influencing only one or the other of true or fraudulent votes would help, as would constraints on any of the effects leading to true or fraudulent votes. Variables like η_7 which respond to only one of the indicatorless concepts but not the other would also help because these would make the error variances on the indicatorless variables have unique implications. η_4's error variance, for example, contributes to the covariance between η_6 and η_7, while η_5's error variance does not.

41. Ways of dealing with this style of colinearity problem are discussed in note 13 to Chapter 6.

42. An author reporting a χ^2 probability of .05 is specifically stating that if the author's model were the true model, we should probably have observed data that were more similar to the model's Σ predictions. If the model were the true model, 19 times out of 20, or with probability .95, data more similar to the model's predictions should have been observed. In *Essentials* (161) I suggested that a χ^2 probability of .2 would be preferable to the usual .05 probability, but experience has taught me stronger respect for the interpretation of .2 as saying that "even if the model were correct, we should have observed more compliant data four times out of five." That is, we are accepting a model when we are still substantially inclined to bet against it! Hence, my recommendation to aim for probabilities in the .75 region before thinking that a model fits the data reasonably well (assuming *N* is under about 500). Eventually this should return to the conventional level of .95 but I recognize that a blind striving for this high a level would prod too many researchers into making unhelpful complementary decisions, so I have tempered my target probability with a dose of realism.

I view as too permissive Anderson and Gerbing's assertion that "in practice, the measurement model may sometimes be judged to provide acceptable fit even though the chi-square value is still statistically significant. This judgment should be supported by ...." (1988:417). The permissiveness arises from the absence of specific criteria being provided for any of the items on the list that followed. This makes it seem as if merely reporting the items on the list is sufficient to allow one to proceed as if the large χ^2 is irrelevant. My experience is that most large-N models with significant χ^2's deserve to fail because they are indeed unacceptable; hence pointing to the largeness of the N and uncritically listing a bunch of other numbers should be insufficient to resurrect them. The emphasis should be on a detailed analysis of the model's diagnostics (most crucially the size of specific residuals the author is willing to overlook or disregard) and the substantive points at issue in the model.

43. Or maybe 5.0 or 6.0, if there are numerous additional identified coefficients and you wish to avoid capitalizing on chance.

44. Recall Section 1.5, and see *Essentials* (119-121).

45. Recall row three of the preface figure, and see *Essentials* (106-116).

Chapter 3

Equivalence, Loops, and Longitudinal Models

Some degree of concern for equivalent models has been around for as long as structural equation modeling itself but interest in the topic was recently rekindled by two articles (Stelzl, 1986, 1991) which provided rules for locating equivalent models, and by the development of the TETRAD program (Glymour, Scheines, Spirtes and Kelly, 1987). TETRAD automated the location of alternative models and came to prominence in the structural equation literature when it seemed to outperform LISREL and EQS at locating proper emendations to some initially misspecified models (Spirtes, Scheines and Glymour, 1990a). In Chapter 4 we will review Stelzl's rules and examine TETRAD to see how it could surpass LISREL and EQS.

In this chapter we focus on the more fundamental issue of why one ought to pay attention to equivalent models. We begin from the idea that equivalent models provide different but equally acceptable interpretations. Reading this in reverse says that one can locate different but equally acceptable interpretations by locating equivalent models. The reversal is important because it provides a lever for extricating oneself from one's entrenched thoughtways. Experience has taught me that my modeling habits are among the most stubborn and devious of my modeling enemies. I expect it may take some effort to persuade you that you are your own worst modeling enemy, and that equivalent models provide some armament in combating the enemy within.

The bulk of this chapter is devoted to examining equivalent models that contain loops. I chose loop models because they provide an opportunity to display how equivalent models can shock us into recognizing and superseding our own modeling habits. At one point these models lead me to defend a *negative* squared multiple correlation as being both reasonable and acceptable. Who on earth would have thought that a negative squared multiple correlation could be reasonable? Not I, or at least not until I pondered an equivalent model.

We will also discuss a model in which a substantial error variance for a latent concept might be viewed as arising from important unlocated causes of that variable, yet where an equivalent model specifically discounts the possibility of an important unlocated cause. Who would have guessed that one's view or interpretation of a model could control whether or not a large error variance is potentially attributable to major unknown causes? Not I, or at least not until I pondered an equivalent model.

We begin with the fundamentals of why equivalent models deserve our attention. This is followed by two definitions of equivalence, and the mathematics that ground an investigation of loop equivalents to a simple recursive model. Even if you skip the mathematics you should be able to appreciate the subsequent demonstration of how loop-equivalent models function and how they provide an open challenge to the interpretation of all effects in recursive models. This chapter ends with two extended examples reporting on equivalents to longitudinal models. The first reports a five-year personal struggle I had with an acceptably fitting model. The discussion of this model is designed to highlight the considerations that led me to be dissatisfied with a well fit model. My struggle ultimately led to the location of a multitude of equivalent well fit models. The second illustration begins with the stability of alienation model from Wheaton et al. (1977). This is the model that leads to both the negative R^2 and the style of theorizing that reconciles large error variances with no need for locating any important missed causes.

3.1 Equivalence and Self-deception

The prototype for all equivalent models is the question of whether a correlation between variables *A* and *B* arises because *A* causes *B*, because *B* causes *A*, because *C* causes both *A* and *B*, or because some more complex structure links *A* and *B*. You will recognize this as the classic discussion that leads to the warning that correlation does not imply causation. Experience with some recalcitrant models that stubbornly refuse to cooperate usually leads to an opposing perspective, namely that the basic modeling task is to locate even one model whose implications are consistent with the observed covariance data, as opposed to selecting one of several acceptable alternatives. Most researchers would be satisfied to have even one model succeed, and overjoyed at finding two models consistent with the available data. From the perspective of a researcher with a stubborn model, the issue of model equivalence seems to be applicable to only those with an embarrassment of riches. But is it an embarrassment of riches, or embarrassment of incapacity when one reports that

A causes *B* and that one has not managed to eliminate the possibility that *B* causes *A*?

Let us begin again, this time starting from the idea that we, you and I, are engaged in more than minimal amounts of self-deception. We are convinced we really know what we are talking about, even though any particular instance of that knowing includes unrecognized biases, sluffs, effective hedges, avoidance tactics, and obfuscating ploys, which keep us from recognizing our own entrapment in a particular thought or modeling style. Our mode of talking, whatever our substantive discipline, is maintained by a paradigmatic outlook that makes us comfortable with our peers yet which appears parochial when viewed in historical retrospect. We label some of our disciplinary contemporaries by the slants or perspectives they hold, but we often fail to recognize that each of us deserves a comparable label, a label which usually only comes into focus in distant retrospect.

I am not suggesting that such entrapment is a sign of personal weakness or moral failing. It is simply unavoidable. But it is not to be uncritically accepted. The dictates of science and academics urge us to actively combat the obviousness that renders our practices seemingly unneeding of analysis. As academics we must seek means of prodding ourselves out of whatever thought-ways have entrapped us, and in whose deceptions we are comfortably yet unjustifiably entwined. The very comfort of the embrace makes it difficult for us to recognize our entrapment, and the comfort renders this the trickiest of our modeling enemies. As academics, we must actively seek tools to pry ourselves out of whatever analytically deceitful embraces we can. From this perspective, the challenge confronting structural equation modeling is to help us see where self-deception has replaced structured argument.

Some people believe that equivalent models are of little concern because one can translate the parameters of one model into parameters of the other equivalent model. I disagree. Though the parameters may be translated, the models may imply that different things are possible or impossible in the real world. The styles of fictions associated with different equivalent models conjure up different visions, and suggest different intervention possibilities. The task is not merely to find a model that fits but to find a fitting model that truthfully mirrors the possibilities inherent in the real world. We are not merely searching for parsimony, as Jagodzinski, Kuhnel and Schmidt (1990) seem to think, but providing some contribution to a substantive field. Making a discipline aware of its own biases and blind spots is about as fundamental a contribution as anyone can make.

It is against this backdrop of the most dangerous enemy within that I read the literature on equivalent models. Failing to pay attention to equivalent models is an admission of insensitivity to one's personal comforts and biases. The biases are blatant if the alternative equivalent models have been articulated, but they remain implicit if the equivalent models are as yet unrecognized.[1] With this said, we are ready to begin our struggle with equivalent models.

3.1.1 Definitions of Equivalence

There are two ways of defining model equivalence, depending on whether one focuses on the discriminating ability of the data at hand or on the discriminating ability of any covariance data that might ever come to hand. The definition of what I will call *broad equivalence* asks us to consider the set of all covariance matrices that are compatible with a particular model specification. Imagine that the locations of the effects have been specified in a model, but that the parameter values are free to vary and hence can lead to a host of different implied covariance matrices.[2] If the sets of covariance matrices producible by varying the values of the coefficients in two different models are perfectly matched, then the two models are said to be broadly equivalent. They are equivalently able to account for the multitude of data sets that might ever be used to test the models. For any possible data covariance matrix, if one of the two models fails, the other will also fail, and if one of the models succeeds at fitting the data, the other also succeeds. The models are indistinguishable because identical patterns of success and failure result from confronting the models with all possible data sets.[3]

Narrow equivalence, or simply *equivalence*, focuses on the ability of two different models to account for a single covariance matrix, namely the covariance matrix **S** the researcher has at hand. If two different models do equally well at reproducing this particular data matrix, the models are said to be equivalent. This is the definition of equivalence we will routinely employ because it corresponds to the problem the researcher typically confronts. One has a set of data, not all the data sets that might ever be derived from a model.[4]

Models that are broadly equivalent must also be narrowly equivalent, though models that are narrowly equivalent need not be broadly equivalent, as illustrated by Luijben (1991).[5] These definitions mainly distinguish the way the equivalence is found, namely through investigation of a particular data set (narrow equivalence) or through mathematical investigation of the similarity of the two models' covariance implications irrespective of the data sets that might be used (broad equivalence as demonstrated mathematically). We will have occasion to use both perspectives, but if we use the term equivalence without qualification, narrow equivalence is intended.

3.2 Loops and Equivalent Models

A substantial portion of *Essentials*, Chapter 8, dealt with the implications and interpretations of models containing loops. It was observed that loops are a special type of indirect effect, and as such, repeated cyclings through loops imply progressively weakening effects. The importance of a loop is that any direct or indirect effects that touch a variable in a loop are enhanced or multiplied by $1/(1\text{-}L)$, where L is the product of the coefficients comprising the loop.[6] Cycling through a positive loop increases the absolute magnitude of any touched basic direct or indirect effect, while cycling through a negative loop decreases any touched basic effect. Sections 3.2.1 and 3.2.2 examine the potential for substituting a loop for specific direct effects in a recursive model.

3.2.1 A Simple Model with a Loop

Figure 3.1 depicts a model containing a loop composed of one observed and modeled variable (η_2) and any number of unobserved/unmodeled variables. This model is underidentified since it contains more coefficients than covariance data points. A covariance matrix for the η's would contain six known values (three variances and three covariances) but there are seven coefficients to estimate in the model: three effects (β_{21}, β_{31}, β_{32}), one loop effect (β_{22}), two error variances (ψ_{22}, ψ_{33}), and the variance of the background cause η_1 (which is ψ_{11} in this all-η representation[7]). The fact that this model is underidentified means that we could not obtain unique estimates of the model's coefficients by using the η covariance matrix as data. But if we knew the values of the seven model coefficients, we would have no difficulty in determining the single η covariance matrix that would be implied by these coefficients. In the next few pages we investigate how the various model coefficients contribute to the covariances among the η's. We will then be able to examine whether particular coefficients can function as replacements for one another, and hence whether broadly equivalent models can be created by interchanging particular coefficients in this model.

We begin by imagining that numerical values have been assigned to each of the seven coefficients in the model in Figure 3.1. We can determine the covariances among the η's that would be implied by this fully specified model by using the equation which represents how η covariances result from the action of the relevant model variables[8]:

$$\begin{bmatrix} \text{Covariances among} \\ \text{the } \eta\text{'s} \end{bmatrix} = (\mathrm{I}\text{-}\mathrm{B})^{-1}(\Gamma\Phi\Gamma' + \Psi)(\mathrm{I}\text{-}\mathrm{B})^{-1\prime}. \qquad 3.1$$

Figure 3.1 A simple loop model.

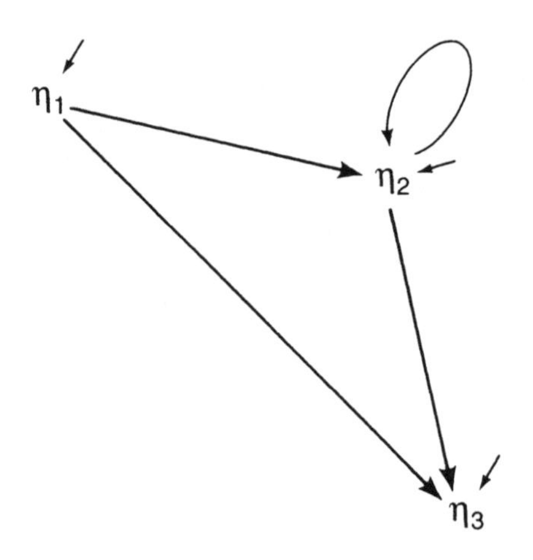

The all-η specification of the Figure 3.1 model means that there are no ξ variables, and hence no Γ or Φ matrices, so the equation informing us of the model-implied variances and covariances of the η's simplifies to

$$\begin{bmatrix} \text{Covariances among} \\ \text{the } \eta\text{'s} \end{bmatrix} = (I-B)^{-1}\Psi(I-B)^{-1\prime}. \quad 3.2$$

The B matrix for the model in Figure 3.1 is

$$B = \begin{bmatrix} 0 & 0 & 0 \\ \beta_{21} & \beta_{22} & 0 \\ \beta_{31} & \beta_{32} & 0 \end{bmatrix} \quad 3.3$$

so

$$(I - B) = \begin{bmatrix} 1 & 0 & 0 \\ -\beta_{21} & 1-\beta_{22} & 0 \\ -\beta_{31} & -\beta_{32} & 1 \end{bmatrix} \quad 3.4$$

and the inverse of (I - B) then is[9]

$$(I - B)^{-1} = \begin{bmatrix} 1.0 & 0 & 0 \\ \frac{\beta_{21}}{1-\beta_{22}} & \frac{1}{1-\beta_{22}} & 0 \\ (\beta_{31} + \frac{\beta_{32}\beta_{21}}{1-\beta_{22}}) & \frac{\beta_{32}}{1-\beta_{22}} & 1.0 \end{bmatrix}. \qquad 3.5$$

By inserting this inverse in the equation for the covariances among the η's (Eq. 3.2), we can write this particular model's implied covariances among the η's as

$$\begin{bmatrix} \text{Covariances among} \\ \text{the } \eta\text{'s} \end{bmatrix} = \qquad 3.6$$

$$\begin{bmatrix} 1.0 & 0 & 0 \\ \frac{\beta_{21}}{1-\beta_{22}} & \frac{1}{1-\beta_{22}} & 0 \\ (\beta_{31} + \frac{\beta_{32}\beta_{21}}{1-\beta_{22}}) & \frac{\beta_{32}}{1-\beta_{22}} & 1.0 \end{bmatrix} \begin{bmatrix} \psi_{11} & & \\ & \psi_{22} & \\ & & \psi_{33} \end{bmatrix} \begin{bmatrix} 1.0 & \frac{\beta_{21}}{1-\beta_{22}} & (\beta_{31} + \frac{\beta_{32}\beta_{21}}{1-\beta_{22}}) \\ 0 & \frac{1}{1-\beta_{22}} & \frac{\beta_{32}}{1-\beta_{22}} \\ 0 & 0 & 1.0 \end{bmatrix}$$

which equals

$$\begin{bmatrix} \mathrm{Var}(\eta_1) & & \text{symmetric} \\ \mathrm{Cov}(\eta_2\eta_1) & \mathrm{Var}(\eta_2) & \\ \mathrm{Cov}(\eta_3\eta_1) & \mathrm{Cov}(\eta_3\eta_2) & \mathrm{Var}(\eta_3) \end{bmatrix} = \tag{3.7}$$

$$\begin{bmatrix} \psi_{11} & & \text{symmetric} \\ \frac{\psi_{11}\beta_{21}}{1-\beta_{22}} & \frac{\psi_{11}\beta_{21}^2}{(1-\beta_{22})^2}+\frac{\psi_{22}}{(1-\beta_{22})^2} & \\ \psi_{11}(\beta_{31}+\frac{\beta_{32}\beta_{21}}{1-\beta_{22}}) & \psi_{11}\beta_{31}(\frac{\beta_{21}}{1-\beta_{22}})+\beta_{32}\mathrm{Var}(\eta_2) & \psi_{11}(\beta_{31}+\frac{\beta_{32}\beta_{21}}{1-\beta_{22}})^2+\psi_{22}(\frac{\beta_{32}}{1-\beta_{22}})^2+\psi_{33} \end{bmatrix}.$$

If we focus on the individual elements in the covariance matrix on the left and the corresponding elements in the right matrix, we see that

$$\mathrm{Var}(\eta_1) = \psi_{11} \tag{3.8}$$

$$\mathrm{Var}(\eta_2) = \frac{\psi_{11}\beta_{21}^2}{(1-\beta_{22})^2} + \frac{\psi_{22}}{(1-\beta_{22})^2} \tag{3.9}$$

$$\mathrm{Var}(\eta_3) = \psi_{11}(\beta_{31} + \frac{\beta_{32}\beta_{21}}{1-\beta_{22}})^2 + \psi_{22}(\frac{\beta_{32}}{1-\beta_{22}})^2 + \psi_{33} \tag{3.10}$$

$$\mathrm{Cov}(\eta_2\eta_1) = \frac{\psi_{11}\beta_{21}}{1-\beta_{22}} \tag{3.11}$$

$$\mathrm{Cov}(\eta_3\eta_1) = \psi_{11}(\beta_{31} + \frac{\beta_{32}\beta_{21}}{1-\beta_{22}}) \tag{3.12}$$

$$\mathrm{Cov}(\eta_3\eta_2) = \psi_{11}\beta_{31}(\frac{\beta_{21}}{1-\beta_{22}}) + \beta_{32}(\frac{\psi_{11}\beta_{21}^2}{(1-\beta_{22})^2} + \frac{\psi_{22}}{(1-\beta_{22})^2}). \tag{3.13}$$

These six equations inform us about how the coefficients in the Figure 3.1 model contribute to the covariances among the η variables, and will serve as our base for examining how the loop functions in Figure 3.1.

3.2.1.1 Investigating the Equivalence of a Recursive and Loop Model

We now consider two models that are special cases of the model in Figure 3.1, and hence whose η variances and covariances are special cases of the six preceding equations. The first model we consider is the *recursive model* which is obtained by setting $\beta_{22} = 0$, and hence whose η covariances may be obtained from the six preceding equations with each appearance of β_{22} replaced by zero. The second model we consider is a *loop model* constructed by retaining β_{22} but eliminating the effect of η_1 on η_3 by setting $\beta_{31} = 0$. The model-implied η variances and covariances for this model is obtained by setting β_{31} to zero in each of the preceding six equations.

Both the recursive and loop models contain six coefficients, and there are six data points, so we may be confronting two just-identified and equivalent models. How can we decide whether the recursive and loop models are broadly equivalent? We must see whether any particular η covariance matrix can be equally well represented as resulting from the parameters in the recursive and loop models. Hence, we will examine each of the covariances on the left of the six preceding equations and enquire into how the parameters in the recursive and loop models would account for this coefficient. This quest will lead us to consider the kinds of connections or relations between the parameters in the two models. If a particular η variance or covariance changed, which coefficients in the recursive and loop models would have to respond if both the models were to continue to match up with the data? Which parameters in the two models function in a parallel fashion in tracking any η covariances changes, as is required if the models are to be broadly equivalent?

To distinguish the parameters from the recursive model from the corresponding parameters in the loop model we will attach an asterisk to the recursive model's coefficients. We now think of the η covariances as specific numerical values and use the six preceding equations to see how the recursive and loop models account for these specific variances and covariances.

From Eq. 3.8 we see that the Var(η_1) equals ψ^*_{11} according to the recursive model, and equals ψ_{11} according to the loop model. Since both ψ^*_{11} and ψ_{11} must equal Var(η_1), they must also equal each other, and we have a direct translation between the first of the six coefficients in the two models ($\psi^*_{11} = \psi_{11}$).

We next examine Cov($\eta_2\eta_1$) and find from Eq. 3.11 that the recursive model accounts for this as $\psi^*_{11}\beta^*_{21}$ since the loop β_{22} is zero in this model. The loop model accounts for Cov($\eta_2\eta_1$) as $\psi_{11}(\beta_{21}/(1\text{-}\beta_{22})$. If the models are equivalent, both models must have identical implied covariances for Cov($\eta_2\eta_1$), and hence

$$\psi^*_{11}\beta^*_{21} = \psi_{11}(\beta_{21}/(1\text{-}\beta_{22})). \qquad 3.14$$

Recalling that ψ^*_{11} must equal ψ_{11}, this simplifies to

$$\beta^*_{21} = \beta_{21}/(1\text{-}\beta_{22}), \qquad 3.15$$

which states that the effect of η_1 on η_2 in the recursive model will equal the corresponding enhanced direct effect in the looped model. The direct effect is

β_{21} and the loop enhances this by 1/(1-L) which in this case is $1/(1-\beta_{22})$.[10] Again we find a relatively simple translation between the coefficients in the two models.

How do the two models account for the Var(η_2)? From Eq. 3.9 we see that the recursive model accounts for this as the left side of the following equation, while the loop model accounts for this as the right side of the equation:

$$\psi^*_{11}\beta^{*2}_{21} + \psi^*_{22} = \frac{\psi_{11}\beta^2_{21}}{(1-\beta_{22})^2} + \frac{\psi_{22}}{(1-\beta_{22})^2} \qquad 3.16$$

Using the previously demonstrated relations between the ψ_{11} variances and the β_{21} effects in the two models, this simplifies to

$$\psi^*_{22} = \psi_{22}(1/(1-\beta_{22}))^2 \qquad 3.17$$

which again is an assertion of the enhancing abilities of the loop. The total or loop-enhanced effect of the error variable on η_2 is $1.0\text{x}(1/(1-\beta_{22}))$, and hence the right side of this equation merely reports that the variance contribution of a variable is the squared structural effect of the causal variable times the variance of that variable.[11] So, we find a third simple translation between the coefficients comprising the two models, where the translation appeals to the enhancements provided by the loop.

We repeat this logic two more times. Next we examine how the two models account for Cov($\eta_3\eta_1$), recalling that there is no loop in the recursive model and no direct effect of η_1 on η_3 in the loop model. After using Eq. 3.12 for the recursive model on the left, the loop model on the right, and simplifying by using the translations already demonstrated above, we conclude that

$$\beta^*_{31} = (\beta_{32} - \beta^*_{32})(\beta_{21}/(1-\beta_{22})). \qquad 3.18$$

Finally, we examine the Cov($\eta_3\eta_2$), do the replacements for the two models in the sixth equation, simplify,[12] and conclude that

$$\beta^*_{32} = \beta_{32} \qquad 3.19$$

which is self-contradictory!

If the effects of η_2 on η_3 in the two models are equal, their difference in Eq. 3.18 is zero, and hence β^*_{31} must be zero. This is contradictory because it demands that the recursive model we postulated as containing three effects really has only two effects. This is another way of saying that if the looped and recursive models are to be equivalent, we cannot permit an effect from η_1 to η_3 in either of the models. If this effect existed in the recursive model, then either Eq. 3.18 or 3.19 must be untrue. It is impossible for both these equations to hold if there is an effect from η_1 to η_3 in the recursive model. Without this effect in the recursive model, we end up with two models containing differing numbers of coefficients: five in the shrunken recursive model and six in the loop model.

So we come to an unexpected conclusion. Though the looped and original recursive models contain the same numbers of parameters, there is no way for the β_{22} loop effect in the loop model to mirror the action of a β^*_{31} effect in the recursive model, and vice versa. Or, synonymously, the models are not broadly equivalent because it is impossible for the coefficients in one of the models to mirror the covariance implications of the coefficients in the other model.

We began by postulating two similar models, where our intention was to delete one standard type of parameter from a recursive model (the β_{31} effect) and replace it with another parameter (the β_{22} loop), that could conceivably have been able to mimic the lost β_{31}. Unfortunately, the β_{22} parameter appears in a location where its value is unable to imply covariance implications that coordinate with the kinds of contributions β^*_{31} makes to the η covariances, or with the constraints a specific β^*_{31} places on the η covariances. That is, β_{22} is not able to replace β^*_{31} by compensating for its action in the model.

It would be incorrect to jump to the conclusion that since β_{22} is unable to compensate for the behavior of this particular effect it will be unable to compensate for all other effects. To mathematically investigate the possibility that the β_{22} loop could replace any of the other coefficients in the Figure 3.1 model, we would have to repeat steps similar to those above for each of the potentially replaceable parameters in the basic model. This would be tedious, so it is fortunate that there is an easier nonmathematical way to proceed, now that we have this particular parameter replacement as a baseline illustrating how a failed replacement behaves. The following section investigates this same attempted replacement using a different procedure. We know the attempt will fail, but what we learn is how the new procedure reports upon a failing replacement. With this in hand, we will be able to efficiently review a host of replacements in the subsequent sections, some of which prove not only to be successful, but also substantively instructive.

3.2.1.2 Illustrating the Nonsubstitutability of β^*_{31} by β_{22}

Let us now consider another perspective on the inability of β_{22} to compensate for β^*_{31} or to substitute for β^*_{31}'s contribution to the recursive model's implications for the η covariances. Imagine that the three η variables in our models each have variance 1.0, and that $\mathrm{Cov}(\eta_1\eta_2)$ is .6, $\mathrm{Cov}(\eta_1\eta_3)$ is .2, and $\mathrm{Cov}(\eta_2\eta_3)$ is .4. Section A of Figure 3.2 presents the estimates one obtains by using this contrived covariance matrix as data, and specifying the recursive model in LISREL.[13] The model is just-identified, and hence the estimates reproduce the covariances perfectly and provide a $\chi^2 = 0.0$ with zero degrees of freedom.

Figure 3.2 A β_{22} loop and how it functions.

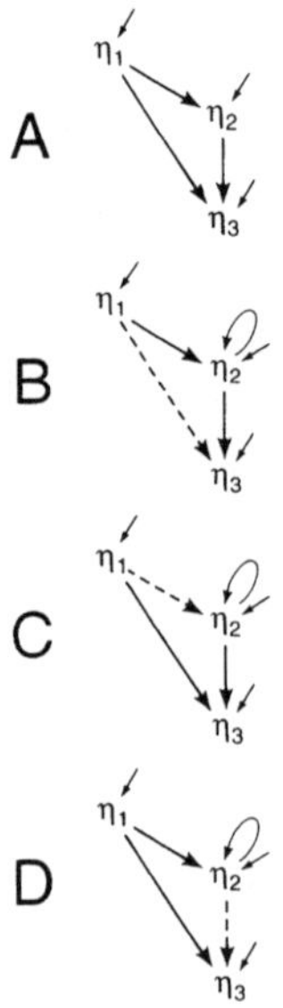

Model		β_{22}	β_{31}	β_{21}	β_{32}	ψ_{11}	ψ_{22}	ψ_{33}	χ^2	d.f.	Estimation Problems MLE
A	A1	.000	−.063	.600	.438	1.00	.640	.837	0.00	0	—
B	B1	−.193	.000	.716	.400	1.00	.911	.840	0.59	0	1,2
	B2	−.193	.100	.716	.460	1.00	.911	.838	0.21	0	1,2
	B3	−.193	−.100	.716	.460	1.00	.911	.838	0.21	0	1,2
	B4	−.193	−.063	.716	.438	1.00	.911	.838	0.00	0	1,2
C	C1	−.500	−.063	.900	.438	1.00	1.440	.838	0.00	0	—
	C2	.000	−.063	.600	.438	1.00	.640	.837	0.00	0	—
	C3	.500	−.063	.300	.438	1.00	.160	.837	0.00	0	—
	C4	.000	−.063	.000	.438	1.00	1.00	.837	88.1	0	1,2
	C5	− 33.6	−.063	−.100	.437	1.00	1194.	.837	89.5	0	1,2
D	D1	−.193	.200	.716	.000	1.00	.911	.960	27.2	0	1,2
	D2	−.193	−.160	.716	.600	1.00	.911	.854	3.98	0	1,2
	D3	−.193	−.063	.716	.438	1.00	.911	.838	0.00	0	1,2
	D4	−.193	.080	.716	.200	1.00	.911	.874	8.40	0	1,2

Note: Fixed coefficients have been shaded.
1 = Warning of a non-identified ψ_{22}.
2 = No standard errors, T-values or correlations among the estimates printed.

The loop model we have been describing appears as row B1 in Figure 3.2. That is, the same "data" matrix is entered, and the value of β_{31} has been fixed at zero, and replaced by the estimated loop effect β_{22}. We observe a direct translation between the estimated effect of η_1 on η_2 in this model and the corresponding estimated effect of η_1 on η_2 in the recursive model. The estimated basic direct effect β_{21} in the loop model is .716, so the loop-enhanced effect is .716(1/(1-(-.193))) = .600 as in the recursive model (recall Eq. 3.15 relating β^*_{21} to β_{21}). Despite this translation, there are signs that this particular loop model is not identified. LISREL reports a warning that ψ_{22} is not identified, and the standard errors, T-values, and correlations among the estimates are not calculable. The non-zero χ^2 with zero degrees of freedom is a clear mathematical impossibility. Hence it is apparent that the loop model is not equivalent to the recursive model, even though some of the parameters in the two models are translatable between the models, as we found above.

Rows B2, B3 and B4 of Figure 3.2 demonstrate conclusively that the estimated value of the β_{22} loop is unable to compensate for the effect of η_1 on η_3. The models in these rows force the effect of η_1 on η_3 to vary between .10 and -.10, but the estimated magnitude of the loop remains unchanged at -.193. Again, all these models display signs of estimation problems, but the problems are not such that they can be cured by manipulating the loop's value. Note also that the estimation problems remain even if β_{31} is fixed at the value obtained as the estimate from the recursive model in row A, namely row B4. In all these models the loop effect is probably underidentified.

We are now thoroughly convinced that the β_{22} loop effect is not a mere translation or reparameterization of the β^*_{31} effect in the recursive model. What is more important, we have seen the kind of diagnostics that report upon this nonequivalence between the models. We will use these diagnostics as a benchmark as we investigate whether this loop might replace any other one of the recursive model's effects, or whether a loop on η_3 would function similarly.

3.2.1.3 Can the β_{22} Loop Replace Other Effects?

We can now "test" to see if β_{22} can replace any of the other effects in the recursive model by entering the alternative specification in LISREL and observing the results. Section D of Figure 3.2 investigates the ability of the β_{22} loop to replace the β^*_{32} effect in the recursive model. The non-zero χ^2's with zero degrees of freedom, and the estimation problems for all of the models in Section D are a convincing demonstration that the β_{22} loop is not a sponge that can absorb the covariance implications of variations in the effect of η_2 on η_3. Clearly the loop model containing β_{22} but omitting β_{32} is not equivalent to the basic recursive model, even if it has the same number of coefficients.

The models in Section C are much more interesting. For three of the five models in this section, β_{22} does seem to be able to unproblematically absorb variations in the effect of η_1 on η_2! Let us begin by examining the two unsuccessful replacement models in rows C4 and C5. These models inform us that when β_{21} is fixed at either 0.0 or -.10 the loop model is unable to function equivalently to the basic recursive model in row A. The reason for these observations is clear. We know that loops "*do not add an effect onto that provided by the basic direct effects, they multiply (or enhance) the basic effects by a factor of the form 1/(1-L)*" (*Essentials*, 261). If the basic direct effect is zero, as it is in the model in row C4, no value assigned to the loop will ever imply an enhanced effect other than zero, because 0.0x1/(1-*L*) remains zero no matter what value the loop, in our case β_{22}, is given.

Similarly, the model depicted in row C5 demonstrates that though a loop can enhance an effect by increasing or decreasing the magnitude of that effect, it is unable to switch the sign of the effect. In the recursive model the effect of η_1 on η_2 is positive. It is fixed at a negative value in the loop model in C5, and the loop β_{22} is unable to undo the improper sign. Positive values of a loop β_{22} imply multiplicative terms $1/(1-\beta_{22})$ that increase the magnitude of the overall effect, while negative values of the loop imply multiplicative terms that decrease the magnitude of the overall effect, but in both instances the sign of the basic effect retains its improper negative value. Figure 3.3 plots the values of 1/(1-*L*) that arise for various values of a loop. Note that the quantity 1/(1-*L*) remains positive for all the values of *L* up to the limiting 1.0 value. These observations inform us why the β_{22} loop in Figure 3.2C5 is unable to reverse, and hence is unable to recover from, a fixed β_{21} effect with an incorrect sign.[14]

Row C2 merely assures us that an unnecessary loop on a recursive model would receive a loop estimate of zero. Fixing β_{21} at precisely the value it has in the recursive model A results in no loop effect being required. It is rows C1 and C3 that carry the bulk of the news about equivalent loop models. Both these models are equivalent to the basic model in that they provide a perfect match to the data, with no signs of estimation difficulties, and they use as many free coefficients as required by the basic recursive model. Both these rows/models postulate different effects of η_1 on η_2 than the recursive model, but they imply exactly the same enhanced effect as the recursive model. That is, $\beta_{21}(1/(1-\beta_{22})) = .9x(1/(1-(-.5)) = .6$ and $.3x(1/(1-.5) = .6$ which matches the effect of η_1 on η_2 in the recursive model. Hence, the loop's value adjusts, via the 1/(1-*L*) enhancement, for the discrepancy between the fixed β_{21} effect in these models and the estimated β^*_{21} effect in the recursive model.

Given the nonequivalence of the recursive model to the loop models in rows C4 and C5, and the equivalence to the recursive model to the loop models in C1 and C3, we are observing what we might call *conditionally equivalent models*. The effect β^*_{21} in the recursive model might be positive or negative, and the loop model with a fixed β_{21} value can be made equivalent to either of these

Figure 3.3 The enhancements associated with various loop strengths.

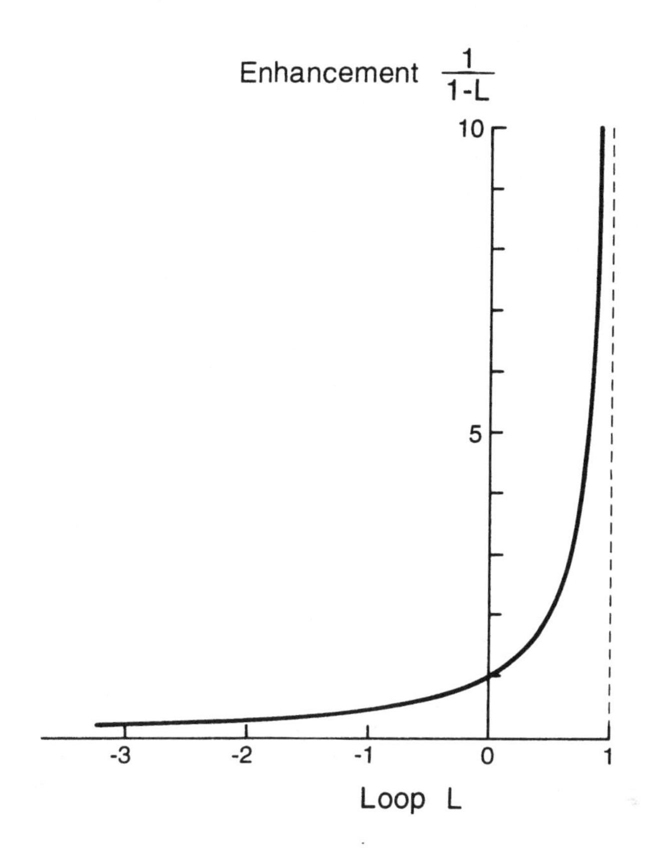

models, with the stipulated condition that the fixed β_{21} value in the loop model must have the same sign as the corresponding effect in the recursive model. The label of *conditional equivalence* also highlights that the equivalent loop models are conditional on a fixed non-zero value of a coefficient, namely β_{21}. Though the recursive model and these loop models contain the same number of *estimated* coefficients, the loop model contains one additional non-zero but fixed coefficient, namely β_{21}. The traditional use of the phrase equivalent models does not distinguish between fixed zero and fixed non-zero coefficients.

It is important to realize that the recursive model and the two equivalent loop models (C1 and C3) do not have exactly the same implications. Consider the effect of η_2 on η_3. β_{32} is estimated as .438 in all three models, but this implies different enhanced effects in the various models. If η_2 were to be increased by one unit, the recursive model predicts that η_3 would respond by increasing .438 of a unit. Contrast this with the loop models where a unit increase in η_2 would result in a series of effects reaching η_3, which would ultimately result in enhanced effects of .438x(1/(1-(-.5)) = .292 and .438x(1/(1-.5) = .876 for the C1 and C3 models respectively.[15]

If the true model was C1 or C3, and we had estimated the equivalent recursive model in row A, our assessment of the impact a unit change in η_2 would have on η_3 could be either an over- or underestimate. Though the models may be equivalent in the sense of being mathematical translations of one another, and though they share the same estimate of β_{32}, the models do not imply the same overall effect of η_2 on η_3. Even with a correct causal sequencing of the effects among the three η's, the recursive model would provide biased estimates of the change in η_3 that would result from a variation in η_2 if either of the equivalent loop models were the true population model.

3.2.1.4 Moving the Loop to η_3

Figure 3.4 uses the same fictional "data" covariance matrix and recursive model as Figure 3.2, but it inquires about the potential equivalence of models having a loop on η_3 instead of on η_2. Hence, we are asking whether the basic recursive model can be thought of as equivalent to other loop models. The results here are very similar to the results that emerged when the loop was attached to η_2. *The only loop models that are equivalent to the basic recursive model are those where the loop estimate can, through its ability to enhance effects, compensate for a fixed non-zero effect of the proper sign that impinges on the variable to which the loop is attached.* The multiplicative enhancements provided by the loop cannot alter fixed zero effects, or reverse incorrectly signed effects. The enhanced effects leading to η_3 in these equivalent models correspond to the effects in the recursive model A. The estimate of the error variance on η_3 fluctuates considerably but always implies a variance contribution to η_3 of about .837, as does the recursive model A.[16] All the models in Section C fail[17] because the β_{33} loop is unable to compensate for an erroneous effect in β_{21}. The β_{21} effect does not touch the loop and hence cannot have its magnitude adjusted by a loop enhancement.

To summarize, *though inserting loops into models can create equivalent models, the loops do not replace any of the effects in the basic recursive model with which one began.* No amount of multiplicative enhancing of a zero effect can recover a non-zero value, and an incorrect sign on an effect cannot be reversed by loop enhancements. The loop must touch a correctly signed fixed effect located precisely where the replaced effect was. *This nonetheless means that all the direct effects in all recursive models can be provided interpretations*

Figure 3.4 A β_{33} loop.

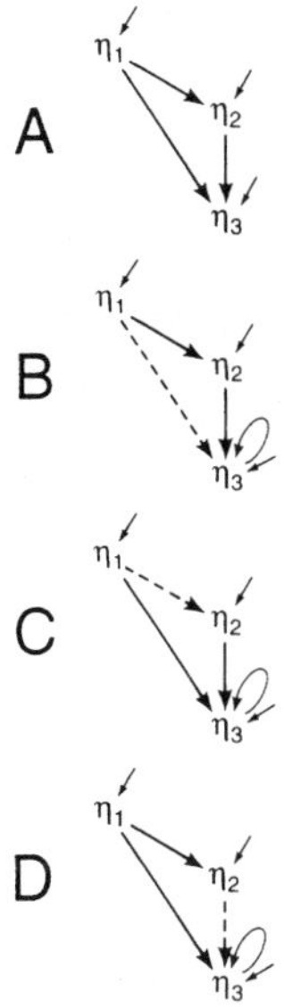

Model		β_{33}	β_{31}	β_{21}	β_{32}	ψ_{11}	ψ_{22}	ψ_{33}	χ^2	d.f.	Estimation Problems MLE
A	A1	.000	−.063	.600	.438	1.00	.640	.837	0.00	0	——
B	B1	−.072	.000	.600	.429	1.00	.640	.965	0.59	0	1,2
	B2	−19.0	.100	.600	7.95	1.00	.640	336.3	0.69	0	1,2
	B3	−.600	−.100	.600	.700	1.00	.640	2.144	0.00	0	——
	B4	−.008	−.063	.600	.441	1.00	.640	.851	0.00	0	——
	B5	.520	−.030	.600	.210	1.00	.640	.193	0.00	0	——
C	C1	−.074	−.067	.900	.470	1.00	.730	.966	26.18	0	1,2
	C2	−.074	−.067	.600	.470	1.00	.640	.966	0.00	0	1,2
	C3	−.074	−.067	.300	.470	1.00	.730	.966	26.18	0	1,2
	C4	−.074	−.067	.000	.470	1.00	1.00	.966	88.81	0	1,2
	C5	−.074	−.067	−.100	.470	1.00	1.13	.966	113.1	0	1,2
D	D1	−.016	.203	.600	.000	1.00	.640	.992	27.17	0	1,2
	D2	−.371	−.086	.600	.600	1.00	.640	1.575	0.00	0	——
	D3	−.001	−.063	.600	.438	1.00	.640	.839	0.00	0	——
	D4	.543	−.029	.600	.200	1.00	.640	.175	0.00	0	——
	D5	−70.9	14.78	.600	−.200	1.00	.640	4996.	27.49	0	1,2

Note: Fixed coefficients have been shaded.
1 = Warning of a non-identified ψ_{33}.
2 = No standard errors, T-values or correlations among the estimates printed.

that demand stronger or weaker (but not null or oppositely signed) basic direct effects. One merely postulates that an unanticipated loop is attached to the appropriate dependent variable. The sign and magnitude of the ignored loop can then be adjusted to defend whatever true basic direct effect one cares to defend. That is, one can claim that any given direct effect in a recursive model is incorrect because the true direct effect is stronger or weaker than the estimated effect. The proponent of the recursive model has not noticed this because a previously unacknowledged loop is attached to the dependent variable, and that loop accounts for the discrepancy between the reported effect and a basic direct effect that would be more to one's liking. The loop-enhanced basic direct effect will equal the magnitude of the original direct effect, but this offers little comfort to those who estimated a recursive model because they did not know how to interpret models containing loops.

From a slightly different perspective, the need for a fixed non-zero effect coefficient seems to place the loop models at a disadvantage. The loop models contain some fixed non-zero coefficients in addition to including the same number of estimated coefficients as the corresponding recursive model. The inability of loops to entirely replace other effects suggests that, in general, some independent justification of the fixed non-zero values in loop models will be required. It would be naive to believe that reviewers will not challenge fixed non-zero effects, so loop models seem to start out with a strike against them.

But the possibility of loops in models should not be ignored. The above demonstrates that ignoring a loop on a variable in an otherwise recursive model can lead to biased assessments of basic direct effects of the variables. Furthermore, this somewhat uninviting beginning for loop-equivalent models can be contrasted to some truly useful applications of loops in the context of longitudinal models.

3.3 Loops in Longitudinal Models

Longitudinal models are models that reflect the specific time sequence in which the data were gathered. Typically, a set of variables is observed at two or more times, with the resulting waves of data being separated by days, years, or decades. The traditional justification for gathering longitudinal data is that this eliminates the possibility of backward, or reverse effects, between the waves. The reverse effects are impossible because a variable occurring later in time cannot cause a variable appearing at an earlier time.

It is standard operating procedure to model the temporally prior value of a variable as a cause of the subsequent value of that variable (e.g., Joreskog and Sorbom, 1988:166-176). The persistence of the variable's values therefore appears as an effect coefficient in the model. Such coefficients are usually called stability coefficients, because an effect of 1.0 (in the presence of no other causes) indicates that the relative values of the variable remained stable over the time interval between the data collection periods.[18] In the presence of other causes, variables typically show less than perfect stability in that the estimates of the earlier observations on the later observation are less than 1.0.[19] The standard way of modeling the longitudinal stability of a variable is depicted at the top of Figure 3.5. The bottom portion of this figure presents an equivalent loop-model representation.

Before we illustrate the estimation and interpretation of this style of loop-equivalent longitudinal model, we should note that one need not wait decades, years, or even weeks to have a longitudinal data set. All that is required is that the observations are sequenced in time. Whenever a researcher makes several replicate measures of an entity, the fact that the measurement procedure is repeatedly applied implies that the measures are sequenced in time, and hence that a longitudinal design may be appropriate. We often hope that the characteristic being measured has remained stable over the course of the replicate measurements, but this is a mere hope, and a hope that can inadvertently obscure interesting features in the data.

3.3.1 An Example: Personal Space Preferences

Permit me a brief biographical sketch. In my youth I was a social psychologist interested in human spatial behavior. I believed that people had differing spatial preferences. Some people were unconcerned by the close proximity of strangers, whereas other people displayed considerable discomfort if a stranger stood closer than arm's length. I did some experiments designed to measure such preferences, and to locate some of the things that might influence spatial preferences: gender, eye contact, threat, direction of approach, and so on. These studies eventually became my Ph.D. dissertation (Hayduk, 1976).

Meanwhile, I was learning LISREL for another research project (Entwisle, Hayduk and Reilly, 1982). While I was writing *Essentials*, it dawned on me that I should try using LISREL on some of my spatial preference data. With all its strengths, LISREL should surely be able to tell me something new about personal space preferences.

The most fundamental issue was that of measurement. I had used the stop-distance technique to measure my subjects' spatial preferences. With this technique, the experimenter slowly walks towards a subject until the subject just begins to feel uncomfortable about the experimenter's closeness. At that point the subject says stop, and the experimenter stops. A measurement is taken of the distance remaining between the experimenter and the subject, using a scale

Figure 3.5 A loop as replacing a longitudinal stability estimate.

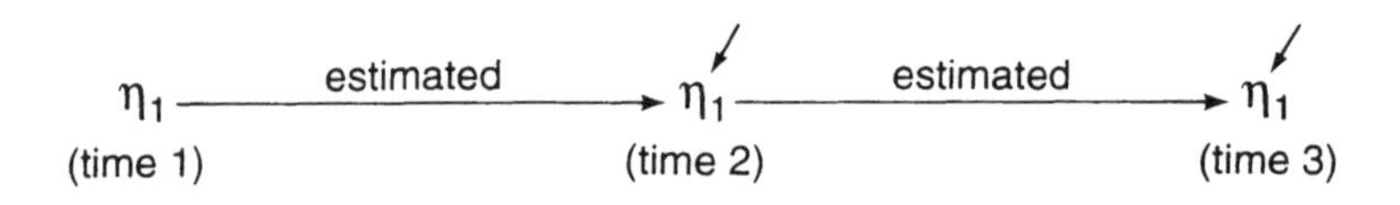

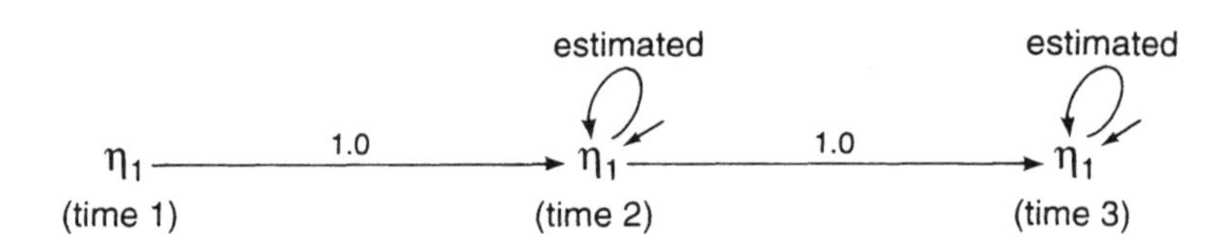

marked on the floor. Even though I had controlled for many things during the measurement process (e.g., eye contact, rate of approach, topic of conversation) this measurement procedure was inherently obtrusive and hence the quality of the measurement was debatable. Here was the LISREL connection. I could use LISREL to estimate the measurement quality, presumably find it excellent, and subsequently use LISREL's χ^2 test to forestall the anticipated critique of the stop-distance measurement technique.

The model seemed obvious. Each person had a personal space preference, and I had multiple measures of that preference imbedded within my experiment. I had one concept (the subject's true spatial preference) with several indicators (ten replicate measurements) of that factor, so the model was a factor model with ten indicators of a single concept or factor (see top of Figure 3.6).

I set up the model, estimated it, and it failed miserably! It had a significant χ^2 and numerous large residuals. Was I to believe that the stop-distance measurements really were not measuring the subjects' personal spatial preferences, and hence that my Ph.D. dissertation deserved to be recalled? After failing to find even a tiny modeling mistake, I resigned myself to the failure of the model and began to think about what was producing the failure.

I reviewed the details of the methodological procedure, whereupon I was struck by the longitudinal nature of the data set. The stop-distance measures had been taken in sequence and there was nothing in the factor model that

Figure 3.6 The factor, simplex, and loop-simplex models (after Hayduk, 1994).

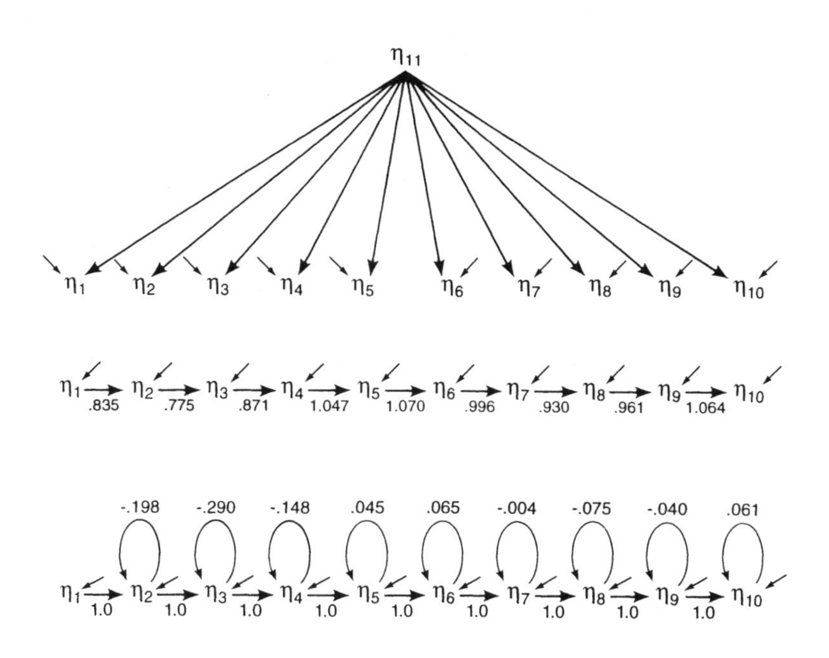

reflected this fact. An alternative, if mildly discomforting, model now became obvious. It was a direct causal chain leading from the first measurement to the second, from the second to the third, and so on along a longitudinal series of observations (see the middle of Figure 3.6). This simple longitudinal model is called a simplex model, and it worked well. The model indicated minimal measurement error variance (between 2 and 3 percent) and the χ^2 was acceptable, with no troubling diagnostics.

The reason the longitudinal simplex model fits is that it correctly predicts that adjacent observations should covary more strongly than observations which are more widely separated from one another. High correlations appear near the diagonal of the correlation matrix, with progressively lower correlations appearing as one moves toward the lower-left corner of the correlation matrix (Hayduk, 1985:143). The factor model does not imply a systematic decline in correlations as one progressively moves away from adjacent observations, and hence it fails to match up with the data.

Then came the psychological crunch. My favored factor model had failed, thereby convincing me that there was no such thing as a personal space preference that was tapped by the ten replicate measurements, yet the simplex model worked and indicated small measurement error. How was I to talk about excellent measurement quality when the factor model had convinced me that there was no "thing" that was repeatedly measured? It was clear that the simplex or longitudinal model carried with it the implication that people's spatial preferences were dynamic in that they changed from moment to moment, but how was one to talk about the shrinking and occasional expanding of personal space implied by the stability or persistence coefficients being slightly smaller or larger than 1.0 (see the middle of Figure 3.6)? The jargon of causation seemed particularly awkward as I somehow needed to talk about the same variable as being both the cause and the effect at each of the segments comprising this model.

I interrupted my writing of *Essentials* long enough to publish the above observations (Hayduk, 1985). I conscientiously avoided saying anything about having measured my subjects' spatial preferences (the factor model had failed) and stressed the inherently dynamic and changing nature of spatial preferences as implied by the longitudinal simplex model.

I remained intrigued by the near contradiction created by the loss of the "thing" or "concept" to be measured, and the "excellent measurement quality" that accompanied the simplex model, which seemed to claim nothing was stable enough to ever get at twice! If repeating a measurement process did not repeatedly measure the same thing, what was the point of saying some "thing" had been measured?

It was the attempt to resolve this tension that led me to a loop equivalent to the longitudinal simplex model (as presented at the bottom of Figure 3.6). I estimated the loop-longitudinal model, fixing each stability coefficient at 1.0 and attaching an estimated loop to each of the endogenous concepts. The models were indeed equivalent. The loop-longitudinal model had the same χ^2, the same minimal residuals, and no signs of estimation problems.

The key point was that the fixed 1.0 values permitted an entity to re-emerge because these 1.0's constitute a claim that is akin to momentum. They claim that "the entity" would persist unchanged were it not for forces beyond the entity itself. The fixed non-zero value did not have to be defended as a selection from among several near competitors. The value 1.0 was dictated by a substantive argument claiming that no change was possible without external intervention. These external forces were of two types: the external random shocks (ζ's) unique to each endogenous time point, and the feedback mechanisms, which were external because they required a causal segment that necessarily carried the effect away from the linear sequence and into some variables in the external world prior to returning any feedback effect at each endogenous time point. The awkward longitudinal model turned out to be equivalent to a model whose essence was the postulation of an entity, or variable, that tended to persist unchanged unless interfered with by forces carried by external factors.

3.3.2 Loops and More Loops

With the concept of a spatial preference now recovered by the perfect persistence, or 1.0 effects, between the adjacent observations, my next task was to characterize the external system that was implicitly invoked through the appeal to loops. Was there really a single unique loop that magically appeared at each of the ten observation times? I was led to investigate a series of loop models, all of which turned out to be equivalent to the basic loop model, and hence also equivalent to the nonlooped longitudinal simplex model!

Figure 3.7 presents just the first segment of these equivalent models. Part A of this figure is a restatement of the single-loop model at the bottom of Figure 3.6. The first variation (Figure 3.7B) proposes that there are not one, but two identical loops attached to the dependent variable at each segment. The requirement that the loops carry equivalent effects implies that there is only one coefficient to be estimated to replace the single loop effect estimated in the basic loop model. This logic extends directly to creating equivalent models by having three or more equal loops replace each of the single original loops (Figure 3.7C).[20]

The next question was, must the loops be attached exactly where the single loop was attached? It seemed obvious that the loop should not double back to include the prior distance measurement, as this would violate the time sequence of the observations. So I postulated that there might have been an intervening personal space preference that had been missed by the measuring process and that this missed preference functioned exactly like the personal space preference that had been measured. In Figure 3.7D the loop at time-2 and the loop on the missed intervening spatial preference (labelled $\eta_{1.5}$) are constrained to be equal, and the error variances at time-2 and the missed intervening time are also constrained to be equal. Again, the model was demonstrably equivalent to the basic loop and longitudinal models.[21]

Two final twists on the replacement of a single loop with other loops are presented in Figure 3.8. In this model each negative loop is conceptualized as a composite of a single initial negative segment that begins the loop effect, followed by multiple equal positive returning effects.[22] The equality constraint was applied only within each set, so there was one coefficient estimate required

Figure 3.7 Alternative equivalent models (first segment only, after Hayduk, 1994).

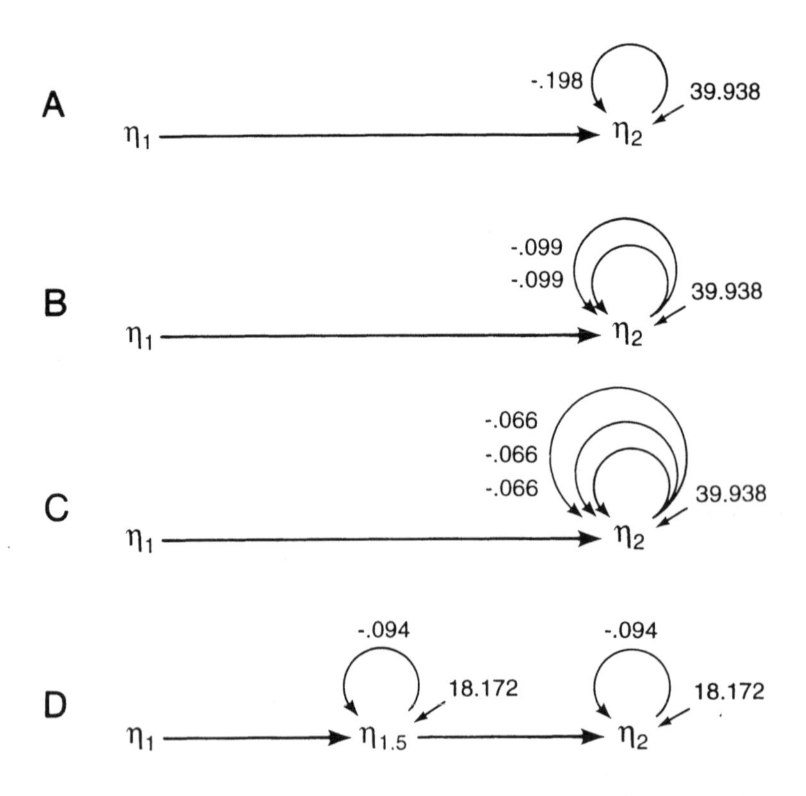

to match up with each of the original single loops. Within each of the sets I chose the number of return effects that would lead to a similar estimate across all of the sets. In this case, I aimed for a value of about 0.40 for all the loop effects and hence I inserted as many multiples of this value as were necessary to approximate the magnitude of the basic loop estimates for the model at the bottom of Figure 3.6. The near but not perfect equivalence of this model arises because this means that the seventh time period receives no loop because the loop effect for this time (in Figure 3.6) was so small that no multiples of .04 were required.[23]

Figure 3.8 Multiple nearly identical loops (after Hayduk, 1994).

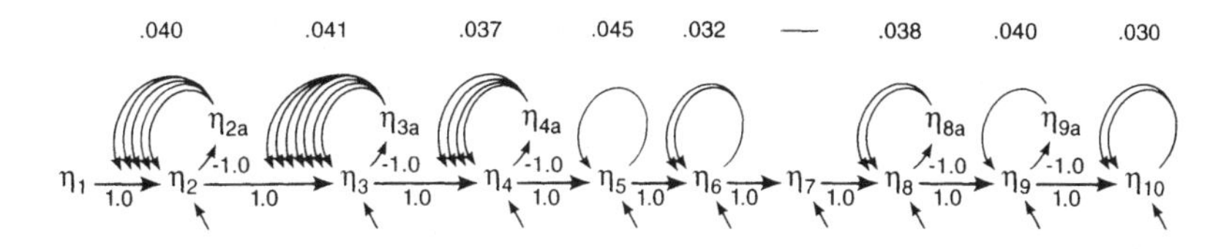

1. The estimated magnitude of each return loop is equal within sets and has the magnitude reported above the set.

What payoff do we get for having gone through the torture of all these equivalent loop models? The reward is that the final model raises an obvious and reasonably compelling analogy. What could be touching a subject's personal space preference such that varying numbers of equivalently effective feedback loops are evoked within the span of a few minutes in a social psychologist's laboratory? Those of you who have been reading Kandel, Schwartz and Jessell's *Principles of Neural Science* (1991) will immediately recognize a parallel to the brain and the equivalence of all axonal transmission potentials.The excitatory and inhibitory neurotransmitters parallel the positive and negative loops, but the key issue is the effect routings, and hence the number of paths via which effects reemerge following various input neural activations. In short, the final model simply looks like personal space preferences are functioning as if they touched various aspects of the human brain at various times.[24] This certainly is not the kind of metaphor that one intuits by staring at the longitudinal model in the middle of Figure 3.6. Hence, the excursion into loop equivalent models led me to a substantive perspective that I otherwise would have overlooked.

A second payoff is that *the above provides a standard alternative interpretation for the basic longitudinal model.* One can interpret the stability coefficients in any longitudinal model as either a sequence of momentary expansions or contractions (effects greater or less than 1.0 in the basic longitudinal model) or as a persistent, momentum-displaying characteristic that touches or activates an ancillary loop or feedback system. The basic longitudinal model leads one to postulate expansions and contractions as the basic components of one's theory, while the loop-longitudinal model leads one to postulate the number and signs of the touched feedback loops as the fundamental theoretical components. The touched side system need not be the brain. It can be any system that responds to and subsequently reinfluences the basic longitudinally measured variable. It might involve peer or family networks, government hierarchies, competing corporations, one's physical body, or feedback effects carried through the atmosphere, depending on the nature of the specific longitudinally measured variable.

Emphasizing touched feedback systems, as opposed to shrinkings and expandings, leads to a substantively different conceptualization of whatever phenomenon is being longitudinally modeled. The equivalence of these models demonstrates that these theoretical styles cannot be distinguished by the covariance data matrix at hand even though the two conceptualization styles focus our attention on radically differing theoretical components.

A third payoff arises because the above begins to develop a whole new approach to estimating and discussing reliability and validity. The implicit model underlying all the classical discussions of reliability and validity is the factor model. In fact, the factor model jargon is so pervasive that we have difficulty imagining any standard alternative. The alternative is now clear. It is the loop representation of the longitudinal model. And it is equally clear that the temporal sequence in which the indicators are observed is a crucial, yet routinely overlooked, piece of evidence.

The basic issue is how to conceptualize stability. Estimating the effect the prior value has on the current value of a variable prods one to view stability as a causal connection between temporally lagged and current values of a single variable. Some external force alters the earlier value of the variable, and whatever causal mechanism transmits the variable from the earlier to the later times intervenes between the times and shrinks or expands the prior value by some estimated amount. In contrast, the equivalent loop model postulates that when some external force alters the value of the basic variable, that variable's new value displays momentum (it persists unchanged indefinitely) until it encounters some external mechanism (at least one other distinct variable) that both responds to the change in the variable's value and subsequently re-influences that variable's value. The persistence is causally effortless, and the variations arise merely because the persisting entity activates an external causal feedback system. We have replaced a model in which temporal persistence is verbalized as self-causing, with a model whose verbalization appeals to natural persistence until the variable causally encounters an outside system which is capable of responding to and subsequently reinfluencing the variable in question.

The loop-longitudinal model does not entirely escape the idea of a variable causing itself, but it does make the important modification that if a variable does causes itself, it is only indirectly through other variables. Any one of the variables in a loop can be thought of as causing itself via a cycle around the loop. A variable indirectly causing itself through the action of other variables is understandable, while a variable directly causing itself merely to carry itself through time seems much more awkward.

Current use of the term "cause" seems to demand that a different variable must appear at the causal end than at the effect end of every causal statement. This requirement seems to locate the gist of the difficulty with understanding how a causal effect can carry a variable through time, as stability coefficients are supposed to do. Without a change from a causal variable to a different effect variable, cause seems to be unintelligible. But with such a change the stability coefficients in a longitudinal causal model seem to demand that the basic variable flits in and out of existence as it changes into something else and then back again as it causally moves from any one time to the next.

The loop stability model, in contrast, renders the issue of persistence noncausal. It merely asserts that the values of variables will persist unless causally acted upon. With this assertion, all the causal action is relegated to an external world where one has interpretive access to other variables as receiving effects. Consequently, one recovers comfortable causal interpretations and avoids the difficulty with having to claim that a variable persists by causing itself, either through some strange sense of cause where the cause and effect are the same variable, or through momentary conversion into other variables and then back again. The difference is whether the causal jargon acts to carry the variable through time, or whether the variable would effortlessly persist unchanged indefinitely were it not for the action of other external variables.

3.3.3 Using Loops to Rethink Another Longitudinal Model

Is what we are learning about loop equivalents generalizable to other longitudinal models? The LISREL 3 through 7 manuals have included the model presented in Figure 3.9, or minor variations thereon, as an example of how to treat longitudinal models. This is a simplification of a model discussed by Wheaton, Muthen, Alwin and Summers (1977) and the covariance matrix for the relevant indicators is provided by Joreskog and Sorbom (1988:169). The basic model examines the stability of the alienation experienced by 932 individuals in rural Illinois who were being confronted by varying degrees of industrial development.

Instead of creating an equivalent loop model by merely replacing the estimated effect of η_1 on η_2 with a 1.0 and adding an estimated loop to η_2, I attempted the more ambitious equivalent longitudinal model depicted in Figure 3.10. The basic alienation concept had been measured in 1967 and again in 1971. Had I modeled alienation71 (alienation in 1971) as being imbedded in a loop, this would have amounted to claiming that there was a potentially complex causal structure in which alienation was imbedded in 1971, but only in

Figure 3.9 The LISREL manual's version of Wheaton et al.'s longitudinal model.

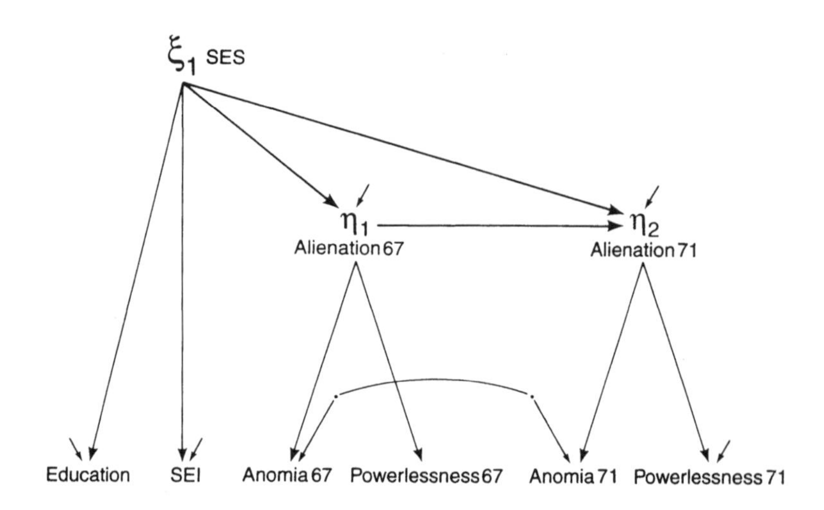

1971. It seemed more reasonable to postulate that not only was there a complex structure in 1971 but that alienation was imbedded in an equally complex structure during each of the intervening years. The Figure 3.10 model postulates that the shrinkage estimated as a β_{21} of .62 in the basic model was due to a complex (loop) process that was operative in each of the intervening years as well as in 1971.

The loop model postulates unmeasured or phantom concepts representing the respondents' alienation during each of the intervening years. The behavior of these unmeasured intervening alienation concepts is modeled in a way which forces them to behave identically to alienation in 1971. The intervening concepts are forced to be as susceptible to error or disturbance terms as is alienation71, and they are required to respond to SES (ξ_1) as strongly as does alienation71, so SES is no more effective in any one year than another. The looped causal structure in which the phantom intervening alienation scores are imbedded is constrained to equal the loop on alienation71 because there

Figure 3.10 An equivalent model with loops and phantom variables.

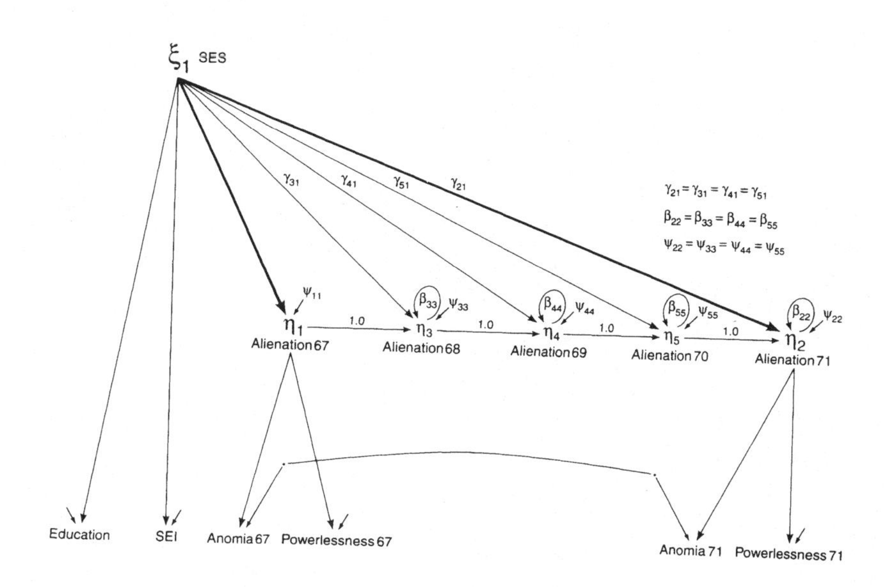

seems to be no reason to believe that alienation was imbedded in any more or less complicated a causal structure during any one of the years.

The loop representation of the Wheaton et al. model contains the same number of coefficients to be estimated as does the original longitudinal model. The estimated stability β_{21} from the original model is replaced by four fixed 1.0 values and a single loop estimate that is entered in four locations (β_{22}, β_{33}, β_{44}, β_{55}). The estimated effect of SES on alienation71 (γ_{21}) is replaced by a single estimated value that is also placed in four locations (γ_{21}, γ_{31}, γ_{41}, γ_{51}).

The loop model is indeed equivalent to the basic longitudinal model.[25] Both models have a $\chi^2 = 6.33$ with 5 degrees of freedom, and no signs of estimation problems. Table 3.1 presents the estimates for the coefficients in the two models. Since all the measurement coefficient estimates (λ, θ_ϵ and θ_δ) are identical in the two models, we will confine our attention to the more interesting structural coefficients, which are influenced by our respecification of the model.

Table 3.1 MLE estimates for the Wheaton et al. and loop-longitudinal models.

Coefficient in		Coefficient Estimate		
Wheaton Model	Loop Model	Wheaton Model	Loop Model	
β_{21}	—	.617	—	
—	β_{31}	—	1.0	fixed
—	β_{43}	—	1.0	fixed
—	β_{54}	—	1.0	fixed
—	β_{25}	—	1.0	fixed
—	β_{22}	—	−.128	equal
—	β_{33}	—	−.128	equal
—	β_{44}	—	−.128	equal
—	β_{55}	—	−.128	equal
γ_{11}	γ_{11}	−.550	−.550	
γ_{21}	—	−.211	—	
—	γ_{21}	—	−.071	equal
—	γ_{31}	—	−.071	equal
—	γ_{41}	—	−.071	equal
—	γ_{51}	—	−.071	equal
ψ_{11}	ψ_{11}	4.705	4.705	
ψ_{22}	—	3.866	—	
—	ψ_{22}	—	1.704	equal
—	ψ_{33}	—	1.704	equal
—	ψ_{44}	—	1.704	equal
—	ψ_{55}	—	1.704	equal
ϕ_{11}	ϕ_{11}	6.880	6.880	

In the loop model the basic indirect effect of η_1 on η_2 (alienation67 on alienation71) is (1.0)(1.0)(1.0)(1.0) or 1.0. This basic indirect effect touches four separate loops, each of which enhances the basic indirect effect. The value of 1/(1-*L*) for each of these loops is 1/(1-(-.128) or .8865, so the enhanced indirect effect of alienation67 on alienation71 is 1.0(.8865)(.8865)(.8865)(.8865) = .6176 which approximates the .617 effect of η_1 on η_2 in Wheaton et al.'s basic model.

What in the loop model is comparable to γ_{21} in the basic Wheaton et al. model (i.e., comparable to the effect of ξ_1 [SES] on η_2 [alienation71] excluding the effect operating through η_1)? The comparable part of the loop model is constituted by one basic direct effect and three basic indirect effects (from ξ_1 through η_5, through η_4 and η_5, and through η_3, η_4, and η_5), each of which is enhanced because it touches one or more loops. The sum of these enhanced direct and indirect effects[26] is -.212 which approximates the corresponding -.211 effect of ξ_1 on η_2 in the Wheaton et al. model. Though the sum of the four equal γ's in the loop model is larger than the -.211 effect of ξ_1 on η_2 in the Wheaton et al. model, the negative loops reduce the basic effects transmitted via the intervening phantom concepts to a value comparable to that observed in the original model.

Similarly, the sum of the four equal error variances exceeds the original error variance on η_2 since the error variance contributions are also reduced because they touch the various loops. One implication of the specification of the four equal error variances is that η_2 appears to be much better explained in the loop model than in the original model. The squared multiple correlations for η_2 are .780 and .501 respectively.

The important difference between the basic longitudinal model and the loop-equivalent longitudinal model is the subtle change in conceptualization between these models. The .617 β_{21} effect in the basic model suggests that the respondents' scores on alienation tend to shrink to roughly 60% of their initial value during the period for 1967 to 1971. In contrast, the negative loop estimate does not mean that alienation tends to shrink, but that any change in alienation is resisted.[27] Alienation would persist unchanged were it not for the loops. Any increase in alienation is resisted, and any decrease or shrinking is also resisted. This is a sign that alienation is stable in the sense that it is imbedded within a system that partially counteracts any disturbances.

Contrast this with the view from the basic alienation model (Figure 3.9) in which alienation seems to display a spontaneous tendency to decay, and hence it has to be continually renewed. The initial alienation score shrinks (to .617 of its initial value) and this shrunken score is supplemented by the SES and random disturbance effects. The loop model suggests alienation neither shrinks nor expands spontaneously, and that the causal network in which it is imbedded makes alienation resist outside influences.

Another, less substantive, difference between the basic and loop-longitudinal models arises in the context of the significance of the structural coefficients. The significance of γ_{21} in the basic nonlooped model is straightforward. T=-4.293 so the effect is significant. As discussed above, this effect from the basic model corresponds to the sum of a series of enhanced direct and indirect effects in the loop-equivalent model. Hence, we should expect that the sum of the corresponding enhanced direct and indirect effects should display a similar significance. We should not expect that the γ_{21} effect in the loop model would display the same T as in the basic model.

Unfortunately, I know of no program that allows the calculation of the significance of the required sum of effects. Since this mixture of direct and indirect effects is not entirely indirect, and is also not quite the total effect of ξ_1 on η_2, since the indirect effect through η_1 is excluded, this implies that LISREL's reports of the significance of total indirect effects are of no help. Clearly, we will have to wait for LISREL, or some other program, to provide for the calculation of the significance of user-specified combinations of effects before the appropriate significance can be calculated in the loop model. For the moment, the only way to determine this significance is to rely on the significance of the equivalent effect in the nonlooped model.

There is a further, and more fundamental question about significance that requires an answer. Namely, should the loop's magnitude be considered in calculating the significance of the effect of ξ_1 on η_2? When I estimated a loop model, which was like the basic nonlooped model, except that it had a fixed 1.0 effect running from η_1 to η_2, and a loop on η_2, the presence of the negative loop increased the magnitude of the effect of ξ_1 on η_2 (namely γ_{21}), compared to the basic non-looped model. This is understandable because the negative loop implies a total effect that is less than the basic direct effect. The mystery is in why the T value for this effect was lowered from -4.293 to -3.488. That is, though the magnitude of the effect estimate increased, the significance of the effect decreased, even though the enhanced direct effect in the loop model was equivalent to the direct effect in the basic model.

Furthermore, the T value for the β_{21} in the basic model (T=12.419) differs from the T value for the loop (β_{22}) in the looped model (T=-4.753).[28] So using the loop to enhance the fixed 1.0 is not judged even approximately as significant as the coefficient replaced by the fixed 1.0 and the loop. Clearly, the current calculation of T values in models with loops leaves a few mysteries for the statisticians to explain.

3.3.3.1 Making Alienation67 Like the Other Alienation Variables

To stretch our loop-equivalent ideas, let us consider whether alienation67 should be modeled in a way that parallels the alienation variables in the subsequent years. We might constrain γ_{11} to equal the other γ's in Figure 3.10, ψ_{11} to equal the other ψ's, and add a loop to η_1 (alienation67) constrained to equal the other loops. Together, these emendations would constitute a claim that alienation67 is imbedded within the same causal network as are the other alienation variables. Though this seems reasonable as far as it goes, it does not go far enough. η_1 is unlike all the other alienation scores in an important respect. It receives no input from an immediately prior alienation variable.[29]

A more reasonable, and much more interesting, extension of the loop model can be created by adding phantom alienation variables (η's) prior to η_1. Figure 3.11 presents a model with eight additional η's (namely η_6 through η_{13}). These new η's and η_1 are constrained to behave similarly in terms of their response to ξ_1 (SES), and to have similar error variances. These new η's behave both like one another, and like η_2 through η_5 in that the loop causal structures in which they are imbedded are constrained to be equal. This model is equivalent to all the prior ways of representing this particular longitudinal model (i.e., same χ^2 of 6.33 with 5 d.f., etc.).

This model challenges us to consider exactly how many prior η's would be required to attain a γ estimate for the prior η's that is equal to the γ estimate for η_2 through η_5. If we really believe our assumption that the effect of SES does not vary, then we know that the model in Figure 3.11 has too few prior η's because a stronger effect from ξ_1 to the prior η's is estimated than to the later η's (-.106 is stronger than -.071). Some number of prior η's would make the two γ estimates equal. For example, yet another equivalent model with 24 prior η's provides a γ estimate of -.074 to those prior η's, which is much closer to the -.071 estimate for SES's effects on alienation in the later years. But even 24 prior years of alienation values is not quite enough.[30]

The researcher is now challenged to ponder what the actual number of prior alienation states (variables) is. And further, if the researcher will not back off the claim that alienation functions consistently during this span of the years, we have located a nonarbitrary way of determining the first, even if unmeasured, annual alienation variable. One merely adds more prior-η's until the two γ estimates are identically equal. If the two γ estimates and the two ψ estimates converge with a specific number of prior η's, we have a nonarbitrary way of locating the beginning of the operation of the longitudinal design.[31] Here we are attempting to answer a question that is not even posed, let alone answered, by the usual way of setting up a longitudinal model (Figure 3.9).

Figure 3.11 An equivalent model with alienation variables in prior years.

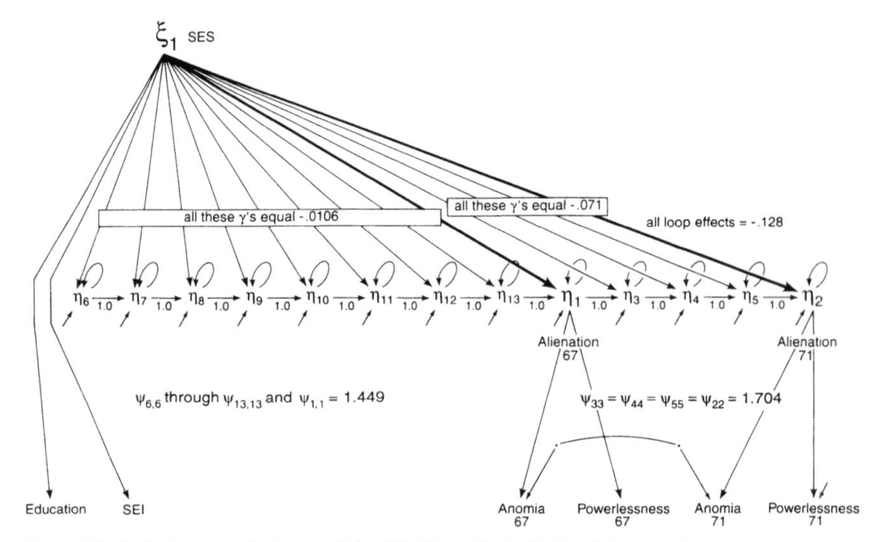

Note: All the λ, θ_ε, θ_δ and ϕ estimates are within .001 of the estimates for the original model.

3.3.3.2 R^2 for Variables Touching Loops

The model with 24 prior alienation variables alerts us to another issue that will eventually require some revision of standard operating practice. The problem concerns the squared multiple correlation, or proportion of explained variance, for variables that touch loops. The result that alerts us to the problem is the -.236 reported squared multiple correlation for η_6, the furthest back of the 24 prior phantom alienation concepts.[32] This -.236 value comes from the calculation of R^2 = 1.0 - (proportion of error variance) = 1.0 - $\psi_{66}/\text{Var}(\eta_6)$ = 1.0 - 1.286/1.040 = -.236. This negative squared multiple correlation is in fact not problematic. It is the traditionally used formula that is problematic.

The formula assumes that all of an error variable's variance is contributed into the dependent variable in an equation. This is not true in general for models with loops. The error variable can contribute more than its variance if the dependent variable touches a positive loop, or less than its variance if the dependent variable touches a negative loop, because the enhancements provided by the loop imply a total effect that differs from the implicit effect coefficient of 1.0 that links the error variable to the dependent variable. In the case of the -.236 R^2 above, the loop effect touching the dependent η_6 is -.128, so the total enhanced effect of the error variable (ζ_6) on the dependent variable (η_6) is 1.0(1/(1-(-.128)) = .8865. The total variance contributed to η_6 from the error variable will be this total effect squared times the variance of the error variable,[33] which is $(.8865)^2 1.286$ = 1.011. Given that the total variance of η_6 is 1.040, the error variable contributes .97 of η_6's variance. This is reasonable in this model because there is almost nothing to explain the furthest back of the alienation variables. η_6 receives only one, now very small, effect from SES.

This suggests that all the squared multiple correlation calculations done by any programs reporting R^2 as "1.0 minus error variance over total variance" are reporting something other than the proportion of explained variance if the dependent variable touches a loop. Usually this will not be obvious because the reported squared multiple correlation will simply be slightly larger or smaller than it in fact should be, and no one will encounter any output which makes them notice any problem with the reported value. The negative R^2 above certainly is noticeable, even though it is not problematic.

It would seem appropriate to calculate the squared multiple correlation for any dependent variable which touches a loop, in a way that acknowledges the loop enhancement to the error variable's contribution to the dependent variable. That is, to calculate and interpret a *Loop Adjusted* R^2, or *LAR*^{2} (pronounced as in **large-square**), as

$$\text{LAR}^2 = 1.0 - \frac{\text{the enhanced error variance contribution of the error variable to the dependent variable}}{\text{total variance in the dependent variable.}} \qquad 3.20$$

With a loop whose effect is L, this will be

$$\text{LAR}^2 = 1.0 - ((1/(1\text{-L}))^2\text{error variance})/(\text{dependent variable's variance}). \qquad 3.21$$

How LAR^2 functions can be illustrated by considering models B3, B4, and B5 in Figure 3.4. In these models the dependent η_3 concept's variance is 1.0 so the ψ_{33} variances for these models of 2.144, .851, and .193 imply R^2's of -1.44, .149 and .807 if one uses the traditional 1.0 minus the proportion of error variance formula. In contrast, if one calculates the enhanced direct effect of the error variable as 1.0x1/(1-L), using the estimated loop effects in these models of -.600, -.008 and .520 respectively, and then calculates the error variable's contribution to the variance of the dependent variable as this effect squared times the variance of the error variable, one finds the error variable contributes .8375 units of variance to η_3 in all the loop models. This is the same error variance contribution as was observed in the basic nonlooped model A1, and so all the models agree that LAR^2 = 1 - .8375/1.00 or .162. Because we had strong negative, near zero, and strong positive loops in the looped models, use of the traditional equation for calculating the proportion of explained variance would, respectively, strongly overestimate, weakly overestimate, or strongly underestimate, the variance contributed to η_3 by the error variable's variance.

The reason the error variable can have much more variance than the dependent variable (η_3) has in model B3 is that the strong negative loop attached to η_3 counteracts any incoming effects, including those effects arriving from the error variable. By ignoring the counteracting effects of the loop the traditional R^2 formula incorrectly assesses the actual contribution the error variable makes to the variance of the dependent variable η_3. Hence, the usual procedure for calculating R^2 will provide an overestimate of the contribution made by the error variable. In general, overestimates or underestimates of the proportion of explained variance will arise whenever the dependent variable touches a positive or negative loop (or reciprocal effect), respectively.

If more than one loop touches a dependent variable, the above formula for LAR^2 is still appropriate if L is the sum of all the loop effects connected to the dependent variable. That is, the error variance implications are the same whether the dependent variable touches one loop with value .6 or three separate loops with values .3, .4 and -.1.[34]

Let us now return to our discussion of loop equivalents to the Wheaton et al. model.

3.3.3.3 Could We Use Smaller Time Segments?

What would happen if many more than three states of alienation were modeled as intervening between alienation67 (η_1) and alienation71 (η_2) in Figure 3.10? Here we are imagining that the loops in the model are operative during time spans shorter than one year. As more intervening phantom variables are introduced, the effects of SES (ξ_1) on each of those variables will decline (eventually approaching zero from the negative side), the loop estimate will approach zero (also from the negative side), and the error variance unique to any given time period will approach zero.

How closely any of these approach zero will depend on exactly how many separate states of alienation are postulated as intervening between the 1967 and 1971 observations. Contrast the 50% error variance for η_2 in the original model with the 32% error variance with three intervening states. There should be less than 1% error if one postulated a new state of alienation at each of the 200+ weekly intervals between 1967 and 1971.

The substantial error variance on η_2 in the original Wheaton et al. model might prompt one to seek other substantive sources of alienation, whereas the minimal error variance unique to η_2 in the loop model with 200 intervening states would suggest that seeking such a variable would be a waste of effort. In the loop conceptualization there would be hundreds of small independent error variances, and nothing large enough to seek or name. The independence of the multiple small error variables dictates that no nameable variable is functioning at the multiple intervening times.

The import of this is that the conceptualizations grounding these statistically equivalent models are now dictating how we should proceed methodologically. According to the basic longitudinal model we should seek to identify major missed causes of η_2. According to the equivalent loop longitudinal model with multiple states of alienation between 1967 and 1971, this would be a waste of energy because this large error is an artifact created by cumulating many small and independent disturbances.

Clearly, the loop representation of the longitudinal model raises a variety of interesting questions which are not even hinted at by the usual representation of the longitudinal model. Each of the numerous equivalent ways of modeling the alienation data represents a different conceptualization of the behavior of the alienation variable. Numerous theoretical decisions and choices remain to be made, and selection from among the alternatives provides some richness and depth to how we think about alienation and to the procedural tactics one might propose in applying for a research grant.

3.3.3.4 Rethinking the Obvious

Let us briefly consider one of the implications of the above. In concluding their discussion of longitudinal models, Wheaton et al. (1977) state what seems to be a near tautology: "Basically we could speculate that stable variables should be less sensitive to situationally based influences, and unstable variables should be more sensitive to these sources of influence." (132). Wheaton et al. are in step with the rest of the literature when they speak of unstable variables as being variables having small stability estimates, such as the β_{21} estimate in Figure 3.9.

Let us reconsider this from the perspective of our loop-equivalent models. If we postulate that the alienation variable exhibits momentum in that its current value remains unchanged unless it is acted on by outside forces, and if alienation is imbedded in a single causal loop touching the later observation (i.e., Figure 3.10 with no phantom intervening η's), what kinds of loop estimate would correspond to a small β_{21}? A strong negative loop effect must be estimated. Specifically, if a loop enhancement of a fixed 1.0 momentum coefficient is to imply an effect that equals .617 in the basic model, we must find that 1.0x(1/(1-loop estimate)) = .617. Solving this equation indicates that the value of the loop estimate would be -.621. Hence, if we are thinking of a longitudinal model with a single touched loop, the loop must carry a strong negative feedback.

Now for the catch. We know that "*negative loops buffer or cushion dependent variables against changes introduced from other variables, and positive loops magnify the responses of the dependent variables they touch*" (*Essentials*, 253, emphasis in original). Hence, with the strong negative loop, we would be observing a system that is well buffered and very resistant to change, or in other words stable because any perturbing forces will be substantially resisted. So the weak persistence in the basic Wheaton et al. formulation, which they view as instability, translates into exactly the opposite, namely exceptional stability and resilience when this is viewed from the perspective of an *equivalent* loop model. Thus, the choice between the basic longitudinal model and the loop-equivalent model leads to radically different assessments of the stability of the alienation variable.

The choice between the models is not merely academic if one is attempting to gauge the likely effectiveness of ameliorative interventions. The loop-longitudinal model implies change will be strongly resisted (there is a strong negative loop). This stands in direct opposition to the claim by Wheaton et al. with which we began, namely that unstable variables should be more sensitive, and hence presumably easily changed. We can now attach some additional depth to Wheaton et al.'s general position that "the stability of a variable over time should be an issue in the theory needed to explain that variable" (1977:132).

I hope this chapter's excursion into loop-equivalent models has convinced you that equivalent models may have some surprising implications. We pursue the issue of equivalent models from a slightly different perspective in the next chapter.

Notes

1. See for example Breckler (1990).

2. Imagine the equation for Σ (Eq. 1.4 in the LISREL manual, or *Essentials* Eq. 4.62) for a particular model, first with symbols designating the various parameters and then with any specific set of numerical values replacing those symbols.

3. This does not mean the models can never be distinguished. Introducing data on means or higher moments, in addition to covariances, or adding more variables to the data set may be able to distinguish between models that are broadly equivalent with respect to covariance data matrices. Hence, even stronger conceptualizations of equivalence are possible. Luijben (1991) uses the broad definition of equivalence in his discussion of equivalence of two models that are expansions of a common base model.

4. For another definition of equivalence, which is based on conditional probabilities and which is used in the TETRAD literature, see Verma and Pearl (1990:220). Note also that these definitions of equivalence could be rephrased to stress the similarity in the two models' implied constraints. It is only when the constraints implicit in the two models are identical that the models are able to predict or account for identical covariance matrices.

5. Luijben (1991:658) points out that equal modification indices typically locate narrowly equivalent expanded models. He also points out that if one is considering expanding a base model, two broadly equivalent expanded models can be created by first locating two coefficients which, if simultaneously freed, make the model underidentified. The expanded broadly equivalent models arise if one frees one of these coefficients and enters a fixed value for the other in one model, and then reverses which coefficient is free and which is fixed for the other model.

6. Or equivalently, L is the proportion of an initial change in any one variable in the loop that returns to that variable at the end of a cycle around the loop.

7. See *Essentials* (209-212) on specifying models with only η concepts.

8. This is the central part of the upper left element in the LISREL manual's Eq. 1.4 (Joreskog and Sorbom, 1988:5) and is discussed in *Essentials* (112-113).

9. This inverse may be obtained via the procedure outlined in *Essentials* (66-67). The steps which progressively convert (I - B) into an identity matrix are as follows. Replace row 2 by row 2 plus β_{21} times row 1 to get a zero in row 2 column 1. Replace row 3 by row 3 plus β_{31} times row 1 to get a zero in row 3 column 1. Divide row 2 by $(1-\beta_{22})$ to get a 1.0 in row 2 column 2. Replace row 3 by row 3 plus β_{32} times row 2 to get a zero in row 3 column 2. The inverse of (I - B) is obtained by multiplying together the four matrices that do these transformations, starting with the first transformation matrix and pre-multiplying this by each succeeding transformation matrix.

10. See *Essentials* (249) for further details on enhanced effects.

11. The reason for the squaring parallels the discussion of *Essentials* Eq. 1.34.

12. The precise steps are: use the sixth equation to write out Cov($\eta_3\eta_2$), adding *'s to denote that these are coefficients in the recursive model, and deleting β_{22} as this is zero in the recursive model. With this as the left side of the equation, now form the right side (the loop models implication for the same covariance) by using the sixth equation again, except that β_{31} drops out since it is zero in the loop model. Next, convert most of the starred terms on the left to unstarred terms using the translation equations developed above. Specifically, $\psi^*_{11} = \psi_{11}$ (in two spots), convert β^*_{21} to unstarred (in two spots) and β^*_{31} to unstarred. Write the term containing (β_{32} - β^*_{32}) as two separate terms, and write the term with ψ_{22} and β^*_{32} as two separate terms. The positive and negative terms with $\beta^*_{32}\psi_{11}$ cancel. A term with $\beta_{32}\psi_{11}$ appears on both sides of the equation and hence these cancel. And finally, ψ_{22} and a squared term also cancel from both sides, leaving $\beta^*_{32} = \beta_{32}$.
13. The y and η variables are equated by specifying single indicators with $\lambda = 1.0$ and $\Theta_\epsilon = 0.0$. The fixed 1.0 λ's scale the η concepts and y indicators to share a common scale or metric. The fixed zero Θ_ϵ's then force the η's to have the same variances and covariances as the input data. N is set at 200.
14. The value of the loop L, or β_{22}, should never reach a value of 1.0 because the quantity 1/(1-L) would become infinite if it did. It might seem that if we used a start value for the loop β_{22} that was greater than 1.0 we could get a reversal in the sign of the effect because 1/(1-L) becomes negative, but this is impossible since a β_{22} greater than 1.0 describes a causal system which is explosive. The effect returning to a variable in the loop would be progressively stronger than the initial variation in the value of that variable. See note 7 on page 274 of *Essentials* regarding loops with values exceeding 1.0.

The inability of a loop to reverse the sign of an effect implies that some models may be conditionally equivalent in that the equivalence holds only for particular values of their coefficients.
15. Recall *Essentials* (252).
16. The total enhanced effect of the error variable on η_3 is 1.0x(1/(1-L)), and it is the square of this enhanced effect times the variance of ψ_{33} that constitutes the overall error contribution of ψ_{33} to η_3's variance. The strong dependence of the error contribution on the value of the loop implies that even in the acceptably fitting equivalent loop models there were very high correlations between the estimates of β_{33} and ψ_{33}. This did not produce any estimation problems, however, because the value of the loop is well determined by the magnitude of the enhancement required to convert the fixed 1.0 effect into the corresponding effect in the recursive model. Once the value of the loop is known, if the total magnitude of the error contribution is to equal that of the basic recursive model, the required error variance estimate is identified by having provided specific values for the estimates of two of the three unknowns in the equation implicit in the first sentence of this note. Hence, even though this equation implies that the values of β_{33} and ψ_{33} are highly correlated, it does not imply that there is any doubt about the estimates of these values. See Section 3.3.3.2 for further discussion of R^2 in the context of loops.
17. The β_{33} loop in model C2 is not identified since using a loop estimate of 0.0 and all other estimates equivalent to those in the recursive model A would also provide a perfect match to the recursive model. That is, β_{33} is not identified in C2, even though the estimates and χ^2 seem acceptable. For this model, it is the notes indicating estimation difficulties that contain the strong evidence of the nonequivalence of this model to the recursive model.
18. The absolute values of the variable may have increased or decreased, depending on the intercept in the equation. The stability of 1.0 merely guarantees that the same rank ordering and absolute inter-case differences are preserved.

19. Even with an effect of 1.0, there can be some change in the values of the variables between observation times if the variance of the variable is increasing due to the effects of other causes. Conversely, an effect of less than 1.0 might mean perfect relative stability of the cases if the variance of the variable is decreasing while no other outside variables have any disruptive causal impacts.

20. One can model multiple loops by creating multiple phantom η's which have no error variables, which each receive an effect of 1.0 from the variable to which the loops are attached, and which return equal estimated effects to the variable to which the loops are attached.

21. Models postulating two or more identical missing intervening preferences are also equivalent and provide progressively weaker loop effects and smaller error variances.

22. In fact, any odd number of negative effects within the loop would do just as well as a single initial negative effect.

23. The basic model has χ^2 = 39.05 with 37 degrees of freedom. This model also has χ^2 = 39.05 but it has 38 degrees of freedom because η_7 received zero multiples of a .04 valued loop. A true equivalence could be attained by choosing a small enough target value so that η_7 would require at least one loop.

24. This could be nearly continual touching if the Figure 3.7D modeling strategy was extended to permit numerous intervening but observationally missed times.

25. LISREL was initially unable to attain reasonable estimates for this loop model because the two-stage least squares procedure LISREL uses to create initial estimates was inserting wild start values at the beginning of the maximum likelihood iterations. This was circumvented by specifying a start value for each of the estimated coefficients (using LISREL's *ST* command) and including *NS* (for no start values) on the LISREL's output line.

26.
$$
\begin{aligned}
-.212 = {} & (-.071)(.8865) \\
& + (-.071)(1)(.8865)(.8865) \\
& + (-.071)(1)(1)(.8865)(.8865)(.8865) \\
& + (-.071)(1)(1)(1)(.8865)(.8865)(.8865)(.8865)
\end{aligned}
$$

27. See *Essentials* (253) for a discussion of negative loops as buffering, cushioning, or stabilizing dependent variables.

28. The switch in sign is understandable, it is the magnitude that is troublesome.

29. In fact, running the model claiming that alienation67 is like any of the other annual alienation scores results in an ill-fitting model with χ^2=135 with 6 d.f. and $p<.000$. The model has one more degree of freedom than the basic model because γ_{11} is not separately estimated. A negative θ_ϵ estimate appears for the powerlessness67 indicator of alienation67. Clearly, the absence of input from an alienation variable prior to alienation67 is more than minimally disruptive.

30. Another way to examine this would be to constrain all the γ values to be equal, and to keep adding time periods until the overall model fit and residuals were unproblematic.

31. The equivalent loop model with 24 prior η's estimated ψ_{11} and ψ_{66} through $\psi_{29,29}$ as 1.286, which contrasts with the estimate of $\psi_{22}=\psi_{33}=\psi_{44}=\psi_{55}=1.704$. This suggests that either there are more disturbances influencing alienation in the later years (which seems reasonable given that the later years spanned a period of industrial development) or that some other aspect of the model (such as the assumption of a consistent loop) needs to be relaxed.

32. This η occupies a position similar to that of η_6 in Figure 3.11 but there are 23 η's, not just 7 η's, intervening between it and η_1.

33. This comes from viewing the equation for η_6 as a special case of the column two model in the preface figure.

34. In terms of computer programming, there is an easier way to proceed than to first locate and then sum all the effects of all the loops touching a dependent variable. One merely goes to the matrix containing the total effects of the η's on the η's and looks for a non-zero diagonal element (call this $TE_{\eta\eta diag}$). Any non-zero diagonal indicates that one or more loops touch this dependent variable (since an effect originates in and ends in this same variable), and the numerical magnitude of that entry is the loop(s)-enhanced effect of the basic loop(s) effect(s) (for example, see the diagonal entries in the top right matrix in *Essentials*, 256). That is,

$$TE_{\eta\eta diag} = L_s(1/(1 - L_s))$$

where L_s is the sum of all the loop effects that touch this dependent variable. Given that $TE_{\eta\eta diag}$ has a known value, L_s is easily calculated as

$$L_s = 1 - 1/(1 + TE_{\eta\eta diag})$$

which is merely a rearrangement of the above formula that arises if one first adds 1.0 to both sides of the equation (as 1.0 on the left and as $(1 - L_s)/(1 - L_s)$ on the right) and simplifies. This value of L_s is the sum of the loop effects, which can now be used in place of L in the formula for LAR^2.

Chapter 4

Equivalent Models: TETRAD and Model Modification

This chapter continues the theme of equivalent models begun in the previous chapter. We review Stelzl's rules for locating equivalent recursive models and note the similarity between these rules and the procedures implemented in the TETRAD program. TETRAD was created by Glymour, Scheines, Spirtes and Kelly (1987) and was initially intended to assist researchers in modifying their current models so that they more closely approximate the true underlying population models. The issue of equivalent models keeps reemerging because two or more of the improved or modified models often turn out to be equivalent.

The logic underlying TETRAD depends upon the same fundamental structural equation modeling ideas used by LISREL, but the way TETRAD employs these fundamental components is quite distinct. It is this alternative perspective on the fundamental thought processes, as much as the recent substantial advances (e.g., Bollen and Ting, 1993), that provides TETRAD its greatest potential.

This chapter acquaints the reader with the style of logic used by TETRAD by placing TETRAD in the context of both structural equation modeling generally, and regression and partial correlations more specifically. The general principles for the comparative evaluation of models that emerge along the way are worth incorporating into one's thinking about models, even though there has been too little use of TETRAD to pass judgement on the program itself. A draft of the TETRAD-II user's manual (Scheines, Spirtes, Glymour and Meek, 1994) indicates that TETRAD is moving toward a modular format with specific subroutine modules designed to address specific research problems. This confuses matters somewhat by permitting different logics to be used in different portions of the program, and by reraising issues of whether multiple indicators are necessary or whether the two-step process should be required routinely.[1] We will overlook these points and focus instead on the general logic of TETRAD and the connection of this logic to structural equation modeling. We conclude by examining the points of match and mismatch between TETRAD, LISREL and EQS as highlighted by the comparative performance of these programs reported by Spirtes, Scheines and Glymour (1990a) and Spirtes, Glymour and Scheines (1993). We begin with Stelzl's rules for locating equivalent recursive models.

4.1 Stelzl on Equivalent Recursive Models

Stelzl (1986) provides four rules for locating models that are equivalent to a base model that is recursive and contains no fixed non-zero effects, and no equality or proportionality constraints.[2] The general argument runs as follows. If one postulates a zero effect in a recursive model, this implies that a particular corresponding zero partial correlation must appear in the data. And, since partial correlations equal specific combinations of the basic bivariate correlations, this demand for a zero partial correlation constrains the possible correlation matrices that can match up with this model. Hence, if one model constrains a partial correlation to zero that is not constrained by another model, some of the covariance matrices compatible with one of the models will not be compatible with the other model, and therefore the models are not broadly equivalent. If two recursive models constrain the same partial correlations to be zero, the correlation matrices are similarly constrained and hence the models are broadly equivalent (Stelzl, 1986:315).

Stelzl examines models that have identical sets of zero effects, and hence identical sets of zero partial correlations, and attempts to describe the changes that can be made to the directions of the non-zero effects in the models without disturbing the zero partial correlations. Switching the direction of an effect in the middle of a recursive model may or may not alter the structural equations in a way which adds or deletes any control variables to a particular partial correlation. Any change in control variables for a zero partial correlation implies a different set of constraints among the basic correlations, and hence nonequivalence of the models.

The key points are: 1) that *in recursive models standardized effect coefficients equal partial correlations*, so inserting a null effect is the same as demanding a zero partial correlation;[3] 2) that *recursive models implying the same set of zero partial correlations are broadly equivalent* from the perspective of the potential covariance/correlation matrices that could ever appear as data; and 3) that *switching the directions of the effects in a model* (by changing the location of the variables in the causal flow) *in ways that preserve the base model's set of implied zero partial correlations results in models that are broadly equivalent* to, or are covariance-indistinguishable from, that base model, even though the estimated magnitudes of the original and reversed effects are different.

Stelzl next observes that a saturated recursive model is equivalent to all the models that can be created by reversing the direction of any one, two, three, or even all the effects in that model.[4] That is, *models saturated with effects lead to a multitude of equivalent models, while sparser models have fewer equivalents.*

The first of Stelzl's four rules arises from considering what happens when a single zero effect is introduced into an otherwise saturated and completely reorderable model. Imagine we start with a saturated recursive model containing variables X_1 to X_7 where the causal sequence or flow is from X_1 to X_7, so X_1 influences all the subsequent variables, and X_7 receives effects from all the prior variables, as in Figure 4.1.[5] In all recursive models, the standardized slope or effect of any one variable on another equals the partial correlation between those variables with all the variables causally prior to the later of the two variables controlled. For example, the standardized effect of X_3 on X_5 equals the partial correlation between X_3 and X_5 with variables X_1, X_2, and X_4 controlled.

If we specify that X_3 has zero effect on X_5, the X_3-X_5 partial correlation must be zero. And, the partial correlation between X_3 and X_5, with the same control variables, must also be implied to be zero by any model that is broadly equivalent to this model. Moving any later variable (X_6 or X_7) to a position in the causal order prior to X_5 (i.e., switching the directions of the several effects connected to one of these variables) results in a nonequivalent model because any such move results in the moved variable becoming an additional control variable in calculating the zero partial correlation between X_3 and X_5. This places a different set of constraints on the basic zero-order correlation matrix from which the partial correlation is calculated, and hence implies that the models are not broadly equivalent.[6] That is, moving any variable that follows X_5 to a causal position prior to X_5 results in a nonequivalent model if any effect leading to X_5 has been postulated to be zero.

One could undetectably reorder the variables X_1 through X_4, or reorder the variables X_6 and X_7 because reorderings confined to just one of these sets or the other leave the partial correlation (standardized slope) at zero. The same set of control variables are used in calculating the partial correlation between X_3 and X_5, no matter which of these reorderings is used. The order in which the variables appear on the list of control variables is irrelevant, just as the order in which the predictor variables appear in a regression equation is irrelevant.

So a nearly saturated recursive model has many equivalent models, but models moving a variable from post-X_5 to pre-X_5 are not equivalent because they

Figure 4.1 Illustrating Stelzl's Basic Procedure.

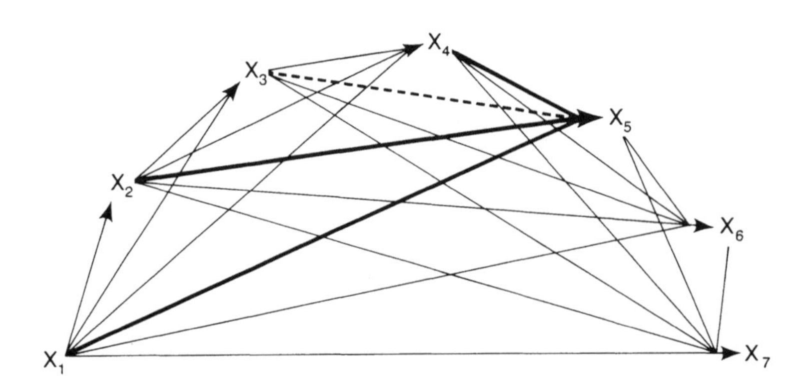

imply that a zero partial correlation arises only if one uses a different set of control variables, and this places a different set of constraints on the bivariate correlations in the data.[7]

Stelzl's Rule 1 is merely the assertion that changes in causal sequence that avoid this later-to-earlier movement of a variable are equivalent because they imply similar constraints on the correlation matrix. Specifically, interchanging the causal positions of (directions of effects among) any of the early variables (X_1 to X_4) produces equivalent models (Stelzl's Rule 1a, 1986:316). Changing the causal positions of (directions of effects among) any of the later variables (X_6 or X_7) produces equivalent models (Stelzl's Rule 1b). Interchanging the positions of just the two focal variables X_3 and X_5 results in an equivalent model (Stelzl's Rule 1c) because the partial correlations of X_5 with X_3 and X_3 with X_5 are identical as long as the same control variables are used.[8]

Stelzl's Rule 2 is a minor relaxation of the general principle that no later variable (X_6 or X_7) is allowed to move to a causal position preceding X_5, the later of the two causally disconnected variables in the scenario outlined above. The relaxation is that an equivalent model arises if X_5 is switched with the very next variable X_6, but only if X_3 has no effect on X_6. That is, if the next variable, but only the next variable, following X_5 also receives no effect from precisely the same variable(s) from which X_5 receives no effects (only X_3 in this case), then X_5 and X_6 could be interchanged undetectably. Again, this is only for the very next variable, and only if that next variable displays precisely the same pattern of absent effects from the prior variables as does X_5. This condition is not satisfied by the model in Figure 4.1. As Stelzl (1986:318-319) explains, these requirements create a special instance when the partial correlation remains the same, despite the additional control variable.

Stelzl's Rules 3 and 4 concern replacing directed effects with error covariances, which are implicitly unobserved common causes. Rule 3 says that if a particular variable, say X_6, receives effects from all prior variables, then any single effect leading into X_6 can undetectably be deleted and replaced by a common cause, that is by a covariance between the error connected to X_6 and the error connected to that prior variable. Just as Rule 2 is a minor relaxation of Rule 1, Rule 4 is a minor relaxation of Rule 3 concerning adjacent variables that happen to have identical sets of excluded causes. Again, under these circumstances replacing a direct cause with a covariance on the errors for the adjacent variables is undetectable.

Though the logic underlying Stelzl's rules is relatively clear, I do not expect that you will be able to confidently apply these rules to even a moderately-sized model after reading the above and studying Stelzl (1986). It would take a master bookkeeper to accurately keep track of a model with ten variables containing multiple zero effects. Furthermore, many of the equivalent models the bookkeeper would locate would be eliminated by appealing to the substantive identity of the variables. Some causal sequencings are simply unlikely for substantive reasons, even though they are not eliminated by the statistical characteristics of the data (cf. Stelzl, 1986:322-329).

If you are not expected to be able to apply these rules, what is expected of you? I would hope that you are now convinced, yet again, that one should seek sparse models, that is, models with many zero effects or other constraints.[9] *The more constraints there are in a model, the fewer the equivalent models there are*. I also expect that you are convinced, yet again, to specifically *develop and estimate alternative models*.[10] The substantively interesting alternative model may just happen to be an equivalent model, and hence equally acceptable from the perspective of your covariance data.

I also expect that you would encourage someone to program these rules so that a computer could do the flawless bookkeeping required to locate equivalent models. TETRAD is nearly that program. It can locate equivalent models, so you should not be surprised when we examine TETRAD in the next section, but it is not exactly "that program" because it works from both tetrads and partial correlations. The connection between tetrads and partial correlations will undoubtedly be opaque until we get to Section 4.2.1.3, so I recommend that you make a mental note to reskim the preceding pages after completing that section. In retrospect, you will also notice that there will have been a shift in focus from "specifically attempting to locate equivalent models" to "improving an ill fit model, which may be done in equivalent ways, and which therefore results in equivalent models." Whether we confront equivalent models after a wilful pursuit, or as an accidental encounter, will eventually pale in importance because more fundamental themes will emerge which span both non-equivalent and equivalent models. As these themes emerge, the issue of equivalence remains; it merely ceases to be the exclusive useful organizing principle. So, let us switch into the context of model modification and try to figure out what tetrads are all about, knowing connections to Stelzl's partial correlations will reappear along the way.

4.2 Model Modification and TETRAD

An article by Spirtes, Scheines and Glymour (1990a) compares the ability of LISREL, EQS, and TETRAD-II to find a true model if the programs are started from a similar, yet misspecified, model. The initial model omitted one or two of the true model's coefficients but had correct starting values for the remaining coefficients. TETRAD-II outperformed both LISREL and EQS. This raised some eyebrows. What was TETRAD doing that LISREL and EQS weren't? We provide an introduction to TETRAD in Sections 4.2.1 and 4.2.2. We return to the Spirtes, Scheines and Glymour (1990a) comparison of LISREL, EQS, and TETRAD in Section 4.2.3 and consider equivalent models more specifically in Section 4.3.

4.2.1 Some Background on Vanishing Tetrads

Neither TETRAD (Glymour et al., 1987) nor TETRAD-II (Spirtes, Scheines and Glymour, 1990a) provides estimates of the coefficients in a model. TETRAD, LISREL and EQS all require that the researcher input a covariance/correlation data matrix, and an initial model specification (location of effects), but TETRAD only attempts to determine the *location* of additional useful effects, it does not estimate any model coefficients. TETRAD attempts to suggest changes (model modifications) that should make a model more consistent with the observed data, without actually calculating the values of the coefficients in either the basic or modified model. This will seem strange to those steeped in the LISREL and EQS traditions. There, a model-implied covariance matrix Σ is calculated from the estimated coefficients, and the residual differences between this Σ and the data ground the modification indices' suggestions for model modifications.[11]

TETRAD begins by examining in turn each and every set of four observed variables that can be selected from the covariance matrix. There are six correlations among any particular set of four variables. Imagine making product-pairs of correlations where all four of the selected variables must be represented somewhere in each pair. There are only three such product-pairs that can be made from the six correlations: $\rho_{12}\rho_{34}$, $\rho_{13}\rho_{24}$, $\rho_{14}\rho_{23}$. TETRAD next asks whether the product of any given correlation pair equals any other pair's product. For example, does

$$\rho_{12}\rho_{34} = \rho_{13}\rho_{24}\ ? \qquad 4.1$$

And/or does

$$\rho_{12}\rho_{34} = \rho_{14}\rho_{23}\ ? \qquad 4.2$$

These equations are called tetrad equations because they are composed of four (tetra) correlations. If both these equalities hold, then the products of the pairs of correlations on the right-hand sides of these equations must also be equal. That is, the tetrad equation created from the two right-hand pairs would also hold. If only one of the above equations holds, then the tetrad equation formed from the pairs on the right-hand sides cannot hold. If neither of these equations holds, the equation made from the right-hand pairs may or may not hold.

Note that the four correlations (two pairs of correlations) in any tetrad equation are not just any four correlations, but are four correlations linked through a focus on four particular variables, and all four variables must be represented on both sides of each tetrad equation.[12]

The tetrad equations may be written using covariances rather than correlations. Using the formula for calculating correlations from covariances,[13] we can express Eq. 4.1 as

$$(\mathrm{Cov}(\eta_1,\eta_2)/\sigma_1\sigma_2)(\mathrm{Cov}(\eta_3,\eta_4)/\sigma_3\sigma_4) = (\mathrm{Cov}(\eta_1,\eta_3)/\sigma_1\sigma_3)(\mathrm{Cov}(\eta_2,\eta_4)/\sigma_2\sigma_4). \qquad 4.3$$

Multiplying both sides of this equation by the product of the four variables' standard deviations cancels out the denominator and provides the covariance **tetrad equation**

$$\mathrm{Cov}(\eta_1,\eta_2)\mathrm{Cov}(\eta_3,\eta_4) = \mathrm{Cov}(\eta_1,\eta_3)\mathrm{Cov}(\eta_2,\eta_4)\ ? \qquad 4.4$$

or the **tetrad difference**

$$\mathrm{Cov}(\eta_1,\eta_2)\mathrm{Cov}(\eta_3,\eta_4) - \mathrm{Cov}(\eta_1,\eta_3)\mathrm{Cov}(\eta_2,\eta_4) = \ ? \qquad 4.5$$

We now turn to the idea that observed covariances are artifacts created through the operation of some underlying causal model and its structural coefficients.[14] Linking the covariance pairs in a specific way therefore also implicitly links sets of the underlying structural coefficients, namely the structural coefficients involved in explaining the covariances of just these four focal variables. This translation to a set of underlying structural coefficients implies that only one of three types of tetrad differences are possible for any given tetrad equation. The structural coefficients comprising the model might:

1) *demand* that there be *zero difference* between the covariance pair products (the products are equal) *no matter what values are assigned to the model coefficients*,

2) *demand* that the covariance pair product difference be *non-zero unless one or more of the model coefficients are zero*, or

3) imply that the difference between the pair products is *indeterminate*, that is, greater than zero, less than zero, or identically zero, depending on the specific estimated values of the structural coefficients.

The first of these points provides a hint of a parallel to Stelzl's arguments. Stelzl argued that a particular effect being zero implied that a specific partial correlation (a particular mixture of basic correlations) must necessarily be zero. Point one broadens Stelzl's argument by not necessarily focusing on merely a single zero effect. Any or all of the models' coefficients might be involved in deriving the tetrad implication, so chains of zero effects or equality constraints might produce the mandatory equality (or zero difference) in point 1 above. And point 1 switches the focus away from the particular mixture of basic correlations called the partial correlation to a slightly different kind of mixture of correlations called the tetrad equations or the tetrad differences.

The second and third points move us beyond Stelzl by demanding that some specific combinations of the basic correlations must be non-zero, and by acknowledging that some combinations of the basic correlations/covariances may be specifically indeterminate.

4.2.1.1 Models Displaying The Various Tetrad Possibilities

We illustrate these possible outcomes by considering the three models in Figure 4.2, which parallel the discussion by Spirtes, Scheines and Glymour (1990a:10-12). Here, we step momentarily out of strict LISREL notation. We are assuming that the variables are observed but since we wish to allow a variety of effects among the variables we refer to them as η's. Hence the effects are β's and the error variables are ζ's. In each of these models we focus on the tetrad equation created from $\mathrm{Cov}(\eta_1,\eta_2)\mathrm{Cov}(\eta_3,\eta_4)$ by switching the inner subscripts to obtain $\mathrm{Cov}(\eta_1,\eta_3)\mathrm{Cov}(\eta_2,\eta_4)$, and hence on the tetrad difference in Eq. 4.5, namely: $\mathrm{Cov}(\eta_1,\eta_2)\mathrm{Cov}(\eta_3,\eta_4) - \mathrm{Cov}(\eta_1,\eta_3)\mathrm{Cov}(\eta_2,\eta_4) = ?$

The factor model[15] in Figure 4.2A implies that the covariance between any pair of variables selected from the bottom row of this model is a spurious

Figure 4.2 Illustrating Tetrad Differences.

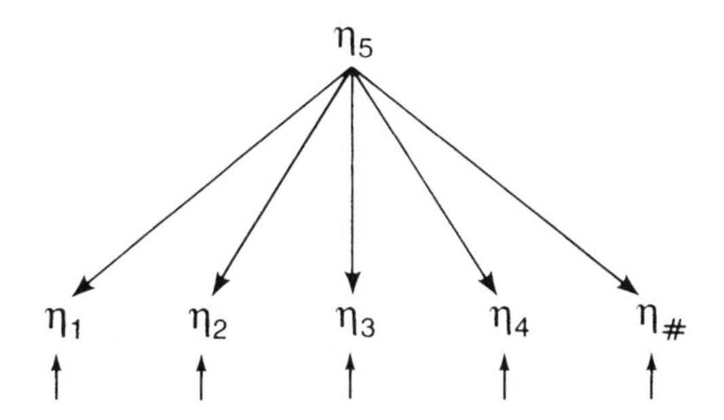

A: A factor model.

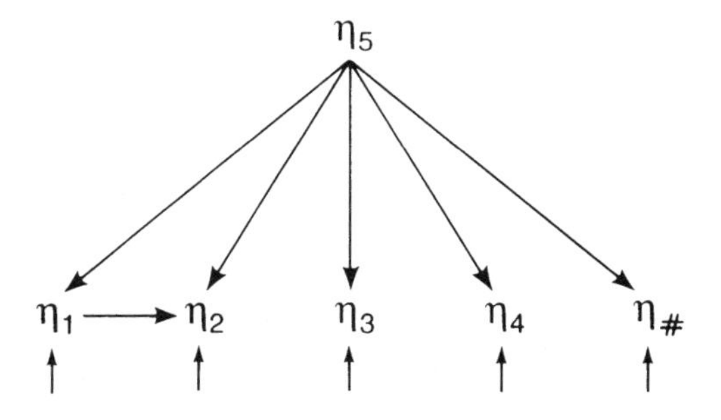

B: One additional effect.

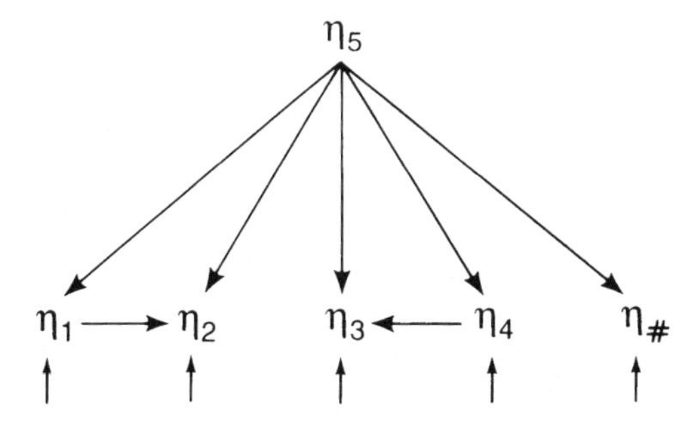

C: Two additional effects.

artifact of the dependence of these variables on the common cause η_5. The covariance provided by a common cause is the product of the variance of the common cause and the structural coefficients leading to each of the pair of dependent variables arising from that common cause.[16] For example, the model implies the covariance between η_1 and η_2 is

$$\mathrm{Cov}(\eta_1,\eta_2) = \beta_{1,5}\beta_{2,5}\mathrm{Var}(\eta_5). \tag{4.6}$$

Representing each of the covariances in corresponding notation, the tetrad difference in Eq. 4.5 becomes

$$[\beta_{1,5}\beta_{2,5}\mathrm{Var}(\eta_5)][\beta_{3,5}\beta_{4,5}\mathrm{Var}(\eta_5)] - [\beta_{1,5}\beta_{3,5}\mathrm{Var}(\eta_5)][\beta_{2,5}\beta_{4,5}\mathrm{Var}(\eta_5)] = \ ? \tag{4.7}$$

which we rearrange to

$$\beta_{1,5}\beta_{2,5}\beta_{3,5}\beta_{4,5}\mathrm{Var}(\eta_5)\mathrm{Var}(\eta_5) - \beta_{1,5}\beta_{2,5}\beta_{3,5}\beta_{4,5}\mathrm{Var}(\eta_5)\mathrm{Var}(\eta_5) = 0 \tag{4.8}$$

which must equal zero because identical terms appear on both sides of the minus sign.

Hence, this particular tetrad difference must be zero *no matter what values are estimated for the variance of* η_5 *or the effects of* η_5 *on* η_1, η_2, η_3 *and* η_4. This model is said to *strongly imply that this tetrad difference is zero* because there are no values for the model's coefficients that could imply anything other than a zero tetrad difference.

Repeating these steps for any set of four variables selected from the bottom row of variables in the model in Figure 4.2A demonstrates that all the tetrad differences created from the bottom row of η's must be precisely zero.

Now consider the model in Figure 4.2B. Here an additional effect from η_1 to η_2 is introduced, but we continue to inquire about the same tetrad difference we examined for the top model, namely

$$\mathrm{Cov}(\eta_1,\eta_2)\mathrm{Cov}(\eta_3,\eta_4) - \mathrm{Cov}(\eta_1,\eta_3)\mathrm{Cov}(\eta_2,\eta_4) = \ ? \tag{4.9}$$

Again, we want to express each of these covariances in terms of how the model accounts for (implies) them. The additional $\beta_{2,1}$ effect is involved in $\mathrm{Cov}(\eta_1,\eta_2)$ and $\mathrm{Cov}(\eta_2,\eta_4)$, but the model implications for $\mathrm{Cov}(\eta_3,\eta_4)$ and $\mathrm{Cov}(\eta_1,\eta_3)$ are similar to those of the top model, since the only connections between these variables are their dependence on their common cause.[17] To stress this similarity to the top model, we first express the tetrad difference as

$$(\mathrm{Cov}(\eta_1,\eta_2))(\beta_{3,5}\beta_{4,5}\mathrm{Var}(\eta_5)) - (\beta_{1,5}\beta_{3,5}\mathrm{Var}(\eta_5))(\mathrm{Cov}(\eta_2,\eta_4)) = \ ? \tag{4.10}$$

in which the covariances altered by the presence of $\beta_{2,1}$ have not yet been expressed as model implications. We now use Duncan's procedure (Section 1.3.3) to obtain $\mathrm{Cov}(\eta_1,\eta_2)$ and $\mathrm{Cov}(\eta_2,\eta_4)$ as implied by the middle model's coefficients:[18]

$$\mathrm{Cov}(\eta_1,\eta_2) = \beta_{2,5}\beta_{1,5}\mathrm{Var}(\eta_5) + \beta_{2,1}(\beta_{1,5})^2\mathrm{Var}(\eta_5) + \beta_{2,1}\mathrm{Var}(\zeta_1) \quad 4.11$$

and

$$\mathrm{Cov}(\eta_2,\eta_4) = \beta_{2,5}\beta_{4,5}\mathrm{Var}(\eta_5) + \beta_{2,1}\beta_{4,5}\beta_{1,5}\mathrm{Var}(\eta_5) \quad 4.12$$

so the tetrad difference in Eq. 4.9 can be expressed as

$$\begin{aligned}&[\beta_{2,5}\beta_{1,5}\mathrm{Var}(\eta_5) + \beta_{2,1}\beta_{1,5}^2\mathrm{Var}(\eta_5) + \beta_{2,1}\mathrm{Var}(\zeta_1)][\beta_{3,5}\beta_{4,5}\mathrm{Var}(\eta_5)] \\ &\quad - [\beta_{1,5}\beta_{3,5}\mathrm{Var}(\eta_5)][\beta_{2,5}\beta_{4,5}\mathrm{Var}(\eta_5) + \beta_{2,1}\beta_{4,5}\beta_{1,5}\mathrm{Var}(\eta_5)] = ?\end{aligned} \quad 4.13$$

Multiplying out the terms in brackets provides

$$\begin{aligned}&\beta_{2,5}\beta_{1,5}\beta_{3,5}\beta_{4,5}\mathrm{Var}(\eta_5)^2 \\ &\quad + \beta_{2,1}\beta_{1,5}^2\beta_{3,5}\beta_{4,5}\mathrm{Var}(\eta_5)^2 \\ &\quad + \beta_{2,1}\beta_{3,5}\beta_{4,5}\mathrm{Var}(\zeta_1)\mathrm{Var}(\eta_5) \\ &\quad - \beta_{2,5}\beta_{1,5}\beta_{3,5}\beta_{4,5}\mathrm{Var}(\eta_5)^2 \\ &\quad - \beta_{2,1}\beta_{1,5}^2\beta_{3,5}\beta_{4,5}\mathrm{Var}(\eta_5)^2 = ?\end{aligned} \quad 4.14$$

from which the first and fourth, and second and fifth terms cancel, leaving the middle term

$$\beta_{2,1}\beta_{3,5}\beta_{4,5}\mathrm{Var}(\zeta_1)\mathrm{Var}(\eta_5) = ? \quad 4.15$$

as the tetrad difference implied by the middle model.

This equation tells us that the middle model implies this tetrad difference can only be zero if one or more of the estimated model coefficients are zero, otherwise it is non-zero. That is, it would take a model modification (elimination of a model coefficient) to make this particular tetrad difference zero if the middle model is the true model. So according to the middle model in Figure 4.2, this tetrad difference must be non-zero. This contrasts with the top model's implication that the tetrad difference must be zero, no matter what coefficient estimates were observed (Eq. 4.8).

Turning to the model in Figure 4.2C, we now enquire about the same tetrad difference, namely

$$\mathrm{Cov}(\eta_1,\eta_2)\mathrm{Cov}(\eta_3,\eta_4) - \mathrm{Cov}(\eta_1,\eta_3)\mathrm{Cov}(\eta_2,\eta_4) = ? \quad 4.16$$

Beginning with the equations for the η's and following the steps parallel to those above demonstrates that this model implies this tetrad equation equals[19]

$$\begin{aligned}
&\beta_{2,5}\beta_{1,5}\beta_{3,4}\mathrm{Var}(\eta_5)\mathrm{Var}(\zeta_4) \\
&\quad + \beta_{2,1}\beta_{1,5}^2\beta_{3,4}\mathrm{Var}(\eta_5)\mathrm{Var}(\zeta_4) \\
&\quad\quad + \beta_{2,1}\beta_{3,5}\beta_{4,5}\mathrm{Var}(\eta_5)\mathrm{Var}(\zeta_1) \\
&\quad\quad\quad + \beta_{2,1}\beta_{3,4}\beta_{4,5}^2\mathrm{Var}(\eta_5)\mathrm{Var}(\zeta_1) \\
&\quad\quad\quad\quad + \beta_{2,1}\beta_{3,4}\mathrm{Var}(\zeta_1)\mathrm{Var}(\zeta_4) = \ ?
\end{aligned} \tag{4.17}$$

This equation has multiple parts that are summed to obtain the tetrad difference, so it is impossible to determine even the sign of this tetrad difference without resorting to an examination of the actual estimated values of the model's coefficients. It would be possible for a zero tetrad difference to arise if some of the structural coefficients were negative, and if the magnitude of the negative products coincidentally cancelled out the positive terms. Hence, the model in Figure 4.2C could be consistent with either a zero or non-zero tetrad difference, but a zero difference would depend on a delicate balancing of the multiple estimates contributing to the tetrad difference. Such a balance would not be inconsistent with the model, but it certainly is not required by the model. That is, the model does not demand a zero tetrad difference for these four covariances, even though specific sets of estimates might just happen to be consistent with a zero tetrad difference.

It should now be clear that the three models in Figure 4.2 imply very different things about the particular tetrad difference we chose to examine. The top model claims it must be zero, the middle model claims it is not zero unless some model coefficient is also zero, and the bottom model might be made to be consistent with a zero, positive or negative difference by a judicious selection of the magnitudes and signs of the model's coefficients.

Spirtes, Scheines and Glymour (1990a) say the top model "strongly implies" that the tetrad difference vanishes because this is an unavoidable consequence of the top model. A non-zero tetrad difference is strongly implied by the middle model (it must be non-zero unless the model is modified by loss of a coefficient), and the bottom model makes no strong implications about this particular tetrad difference.

4.2.1.2 Using Tetrad Differences as a Criterion

Now we encounter an idea that moves the discussion well beyond Stelzl's conceptualization. Recall that in the current notation, the η variables are observed variables so that the covariances comprising the tetrad differences above are observed quantities. *If the observed tetrad difference turns out to be zero, would the three models be equally well supported?* Clearly they would not. The top model would be most strongly supported because it predicted that it could not be otherwise. The middle model is contradicted because to account for the zero tetrad difference the model must be modified by deleting at least one model coefficient (making it zero). The bottom model might be brought into accord with the observation of a zero tetrad, but only with a judicious selection of values for the model coefficients.

Note specifically that we do not need to estimate the coefficients in any of these models to arrive at these comparative decisions. We need the form of each model to get the symbolic representation of the tetrad implications as in Eqs. 4.8, 4.15 and 4.17, and we need the data covariances to calculate the actual value of this particular tetrad difference (Eq. 4.5), but we do not need to estimate any β or other model coefficients to arrive at a comparative assessment of these models.

If a zero tetrad difference is in fact observed, two evaluative criteria are involved in rating the top model as most strongly supported and the middle model as least supported. The middle model is problematic because it implies something that is inconsistent with the data (we observe a zero tetrad difference despite the model's implication of a non-zero difference). Glymour et al. call this *Thurstone's principle* (or the falsification principle) and summarize it as a claim that "a model should not imply constraints that are not supported by the sample data" (1987:94). They refer to the principle that makes the top model preferable to the bottom model, given the assumed null tetrad difference, as *Spearman's principle* (or the explanatory principle)[20] and state it as follows: "Prefer those models that, for all values of their free parameters, entail the constraints judged to hold in the population" (1987:94).

Thurstone's principle is widely appreciated and parallels the logic underlying the usual χ^2 test of the difference between the data covariance matrix (S) and the estimate-and-model implied covariance matrix (Σ). It is the dedication to Spearman's principle that seems new. Spearman's principle is not merely a restatement that parsimonious models (models with few coefficients) are preferred.[21] It is a recognition that the top model deserves credit for boldly asserting that this data characteristic (the zero tetrad difference) could not be otherwise, while the bottom model equivocates with "it depends on the values assigned to the various coefficients." Those accustomed to seeing regression slopes change as new predictor variables are entered[22] will be uncomfortable with the bottom model's capitalizing on a delicate, and potentially unstable, balancing of multiple coefficients' values to match with the putative zero tetrad difference.

4.2.1.3 Triad and Pentad Differences

Our discussion so far has focused on tetrad differences, but a similar logic holds for partial (triad or three-variable) and pentad (five-variable) differences. Had we begun by examining the correlations among three, instead of four, variables, we would have found that models can differ regarding their predictions for partial correlations in the same way that models can differ regarding tetrad differences. A model can strongly imply a zero partial correlation (a zero partial correlation is implied no matter what values the model's coefficients take on), or be made to be consistent with a zero partial correlation by a delicate balancing of multiple coefficients' estimates, or be unable to account for a zero partial correlation without entering a mandatory zero value for one or more of the model's coefficients. Do you see Stelzl's ideas as being subsumed within this?

To clarify this a bit, consider the usual formula for the first order[23] partial correlation between two variables (η_1 and η_2) controlling for η_3:

$$\rho_{12.3} = \frac{\rho_{12} - \rho_{13}\rho_{23}}{\sqrt{(1-\rho_{13}^2)}\ \sqrt{(1-\rho_{23}^2)}} \ . \qquad 4.18$$

To make the numerator and hence the partial correlation zero, there would have to be no difference between ρ_{12} and the product of ρ_{13} and ρ_{23}. Paralleling Eq. 4.5, we could express this as a triad difference: $\rho_{12} - \rho_{13}\rho_{23} = ?$

Some models may imply this partial or triad difference is zero, no matter what coefficient values are used (e.g., the model in Figure 4.2A if η_5 is the control or "third" variable linked to a correlation between η_1 and η_2)[24] while other models imply this difference will not be zero unless some specific coefficient is zero (e.g., β_{21} in the middle and bottom models in Figure 4.2, if we are using the correlation between η_1 and η_2 and controlling η_5).[25] Still other models might be able to account for this with a delicate balancing of coefficients. Hence, had we observed a zero triad difference, we could use Thurstone's and Spearman's principles to decide how much credit to give each model, based on what the model predicts for this particular partial difference, parallel to how we handled tetrad differences. Furthermore, nothing but the need for a combinatorial increase in computing power seems to be stopping us from considering pentad (five covariances) or higher order differences in a parallel manner (cf. Glymour et al., 1987:85).

No one has yet demonstrated that LISREL or EQS is sensitive to the patterns of implied zero tetrad, triad, or pentad differences, so these patterns among the data covariances seem to provide additional information for assessing a model's adequacy beyond assessments based solely on the discrepancies between elements of S and Σ.

4.2.2 The Logic of TETRAD and TETRAD-II

We can now characterize how the TETRAD program functions. One inputs a data (correlation or covariance) matrix and base (small) model. The program scans the data matrix, looking in turn at each set of four variables and creating and recording the observed tetrad differences. Each difference is tested to see if it is statistically significant,[26] so it is clear whether the data provides evidence for or against a zero difference. Similarly, all the possible partial or triad differences are created and tested to see if they differ from zero.

The program then turns to the model and expresses each tetrad and triad difference as a function of the model's coefficients. This demands that TETRAD do operations parallel to the steps we did in deriving Eqs. 4.8, 4.15 and 4.17. This is a formidable programming task. TETRAD must be prepared to do this for any model, no matter what the base model might look like.[27] By building equations similar to 4.14 and seeing which coefficient sets cancel out, TETRAD determines whether the model strongly implies a zero difference, implies a non-zero difference, or is noncommittal about the difference (a delicate balance).

TETRAD reports the results of these calculations, but more importantly it provides diagnostics that indicate how these calculations would change if particular effects were added to the base model. The key diagnostics concern how a particular potentially includable effect influences the model's ability to satisfy Spearman's and Thurstone's principles. If including a particular effect reduces a model's ability to strongly imply tetrad differences that are zero in the data, this is a serious loss from Spearman's perspective. If including the effect allows the model to imply a non-zero tetrad difference that is in fact non-zero according to the data, this is a gain from Thurstone's perspective.

The multiplicity of implied changes in tetrad and partial differences, and the multiplicity of potential patternings in the data mean numerous conditions must be weighed against one another to decide whether any given potential effect should be included or not. These decisions were user determined in TETRAD but became automated in TETRAD-II through the inclusion of a function specifying the relative seriousness of violating Spearman's and Thurstone's principles. Hence, TETRAD-II can suggest model revisions (effects, or common causes to include) which provide minimum violation of the various principles. Any given base model may have several acceptable first revisions, and for each of these revisions TETRAD-II cycles through the options for suggested second revisions, then suggested third revisions, and so on, each time trying to locate the effect whose inclusion provides minimal disruption of Spearman's (attainment of justified assertiveness), Thurstone's (avoidance of unjustified assertiveness), and general parsimony principles.[28]

It takes considerable effort to plough through a model's equations to see if adding a particular coefficient changes the model's implications for even one tetrad difference, and it requires considerable brute force to examine the impact of even one possible change on each and every tetrad difference that can be formed from the variables in a particular data set. TETRAD has been programmed to be as efficient as possible,[29] but the program still requires about as much brute computing power as does LISREL or EQS.

TETRAD users must make numerous technical decisions, including the level of significance that is required to claim a tetrad difference does/does not hold in the data,[30] and the relative weight or importance to assign to violations of Spearman's and Thurstone's principles. Default values (decision criteria) are provided in TETRAD-II, but this program is too new to have accumulated any substantial assurance that the defaults are reasonable in particular research domains, or that they are robust to violations, which are likely if the default conditions are used without knowledgeable examination.

And there is a dash of mystery. Spirtes, Glymour and Scheines (1991a:21) report that "a procedure of this kind has been implemented in the TETRAD-II program." The mystery arises because "this kind" does not refer to the full-blown discussion of model-implied tetrad differences (as above), but to two theorems which use conditional probabilities as a way to locate models.[31] That is, TETRAD-II does not use precisely the same procedures as did TETRAD. It tries to accomplish the same thing with a slightly different, and presumably more efficient, algorithm.

As much as we might appreciate the additional efficiency, the switch in procedures means we will have to rebuild the basics. Fortunately, TETRAD-II's basics are relatively easily summarized. TETRAD-II begins with the observed covariance matrix, and first determines the location of all direct effects that will ultimately be required in the model. This is done by checking to see if a pair of variables is ever uncorrelated if one controls for any or all of the other variables in the data set. That is, TETRAD-II selects two focal variables and examines the first to highest order partial correlations that can be made using all the other variables in the data set. If the two focal variables are never uncorrelated under any set of available control variables, a path or "edge" is required between those variables, though the direction of that path remains unknown. This is Spirtes, Glymour and Scheines' (1991a:9) Theorem 1, and it resembles Stelzl's use of partial correlations, but it looks at the partial correlations between each pair of variables controlling for any or all the other variables in the data set, not just the causally prior variables in a recursive model.

Spirtes, Glymour and Scheines (1991a:9) then provide a direction to the set of edges (located paths with as yet undetermined directions) which were determined on the basis of examining the multiple partial correlations for each possible pair of variables in the data matrix, as described just above. TETRAD-II does this by focusing on a set of three variables at a time, where the three variables have been selected such that one of the three has edges linking it to the other two, but the other two variables are not directly connected by an edge (see Figure 4.3).

The possibility in Figure 4.3A is excluded by the condition that only two of the three edges are present. Spirtes, Glymour and Scheines' "second theorem" allows TETRAD-II to decide among the possible options for the direction of η_2's edges (depicted as B through E in Figure 4.3) by investigating another data-provided condition. Figures 4.3B, 4.3C, and 4.3D share the property that variables η_1 and η_3 become statistically independent if one controls for the value of variable η_2. This controlling eliminates the covariance contributed by a common cause (4.3D) or eliminates the covariance arising from a connecting mechanism (4.3B or 4.3C), and hence the variables become independent.

If the variables η_1 and η_3 are not uncorrelated with just η_2 controlled, then the only acceptable causal ordering is that the causal arrows are pointing toward η_2 (Figure 4.3E), where the lack of independence merely reports on the fact that the two causes of η_2 are correlated, even though the absence of an edge between these variables indicates that this correlation must be due to the action of other variables in the model. This procedure is now repeated for all combinations of three variables satisfying the condition of "two edges focused on one variable." The process keeps getting easier, as some of the previously determined directions to the edges assist in determining the direction to the edges of other triplicates of variables, which are now possibly mixtures of mere edges and directed paths.

Thus, in two steps, based on first a full screening of conditional independence/dependence and then repeated examination of particular conditional independence, one moves from no model whatever to a model composed largely, even if not necessarily entirely, of directed effects.[32] This technique is not foolproof,[33] but this is not what concerns us here. The point is that this conditional probability based justification differs from the TETRAD procedure of implicit comparison of sets of structural coefficients. Actually, the difference is not as big as it might initially seem. The formulas for partial correlations of first or higher orders of controlling that are used to determine conditional independence contain exclusively covariance information on the right hand side. Hence, if structural coefficients give us covariances, and covariances give us conditional independence/dependence there is more than minimal connection between the procedures.

Clearly, TETRAD-II's programming was, and still is, in flux. A draft version of a user's manual for TETRAD-II (Scheines et al., 1994) indicates that the program will now be implemented as a set of modules or subroutines, some of which use tetrad differences, others of which use conditional probabilities, and still others of which use both tetrads and conditional probabilities and/or recent innovations for solving context specific problems.

TETRAD has a lot in its favor. It capitalizes on an aspect of modeling that is currently underutilized. Specifically, tetrad equations and differences are grounded in *patterns* in the covariance matrix, and the diagnostic worth that can be wrung from these patterns. Traditional recommendations to construct parsimonious models (Duncan, 1975) are likely to improve a model's chances

Figure 4.3 Tetrad-II's Other Logic.

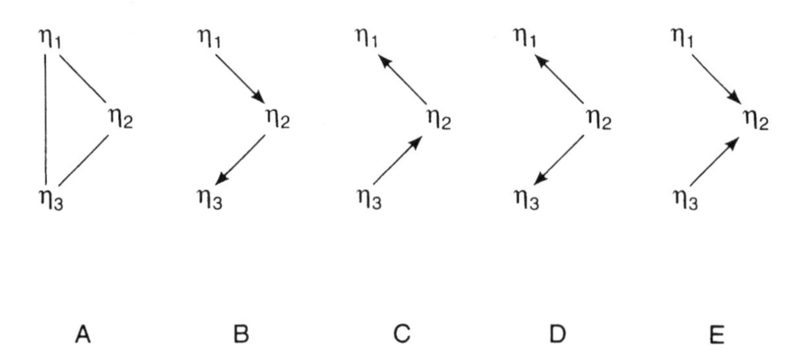

of satisfying Spearman's principle (prefer models that strongly imply observed patterns), but there has been no way to consistently implement Spearman's principle in the evaluation of models. Another strength is TETRAD's emphasis on finding the implications of the *placement* of coefficients as opposed to the magnitudes of these coefficients.

TETRAD's ability to start, and preference for starting, from simple models is both a strength and weakness. We can laud it for demanding less theorizing, but fault it if it encourages less theorizing. The attempt to build a model from purely data-driven constraints makes TETRAD-II, or at least TETRAD-II's Build and MIMBuild modules, more ambitious than TETRAD, which started from a substantial base model. This implicitly reduces or eliminates the role played by theory in model construction. The only place theory seems to be required is where the data-driven procedures for locating and then directing edges fail to be sufficient to specify directed effects. That is, if one starts from a null model, then theory enters the picture precisely and only where one has moved beyond what is decidable by the data. The data is not a test of the theory and the theory is always subservient to data in the sense that the theory need not make any assertions about effects prior to the data having had its full say. From this perspective, the extra ambitiousness resulting from TETRAD-II's attempt to start from a null model is a step toward data dredging, even if reasonable and responsible data dredging, and away from examining the implications or adequacy of specific theories. Fortunately, there are other modules within TETRAD-II (Tetrads and Search) that assume one is starting from a substantial model and hence which are much more theory oriented.

Another positive is that there is an active cadre of proponents of TETRAD working on programming improvements and spreading the news (Glymour et al., 1987, 1988; Glymour, Scheines and Spirtes, 1988; Spirtes, Glymour and Scheines, 1991a, 1991b, 1993; Scheines, Spirtes, Glymour and Meek, 1994; Glymour and Spirtes, 1988; Glymour, Spirtes and Scheines 1990, 1991a, 1991b; Spirtes, 1991; Spirtes and Glymour, 1991; Scheines and Spirtes, 1992; Spirtes et al., 1991).

The negative side of the tally finds that the most substantial relevant works on TETRAD and TETRAD-II (Glymour et al., 1987; and Spirtes, Glymour and Scheines, 1993) are reader unfriendly. These works debate numerous side issues, and do not have a logical sequence that leads, let alone forces, a novice reader through discussions of reasonable ways of making the key user decisions. Nor do they attempt to highlight pitfalls or misuses. The examples in these books clearly indicate that the program's authors can use TETRAD effectively, but little attention has been given to turning new acquaintances into proficient users. The recent manual (Scheines et al., 1994) is by far the easiest of the major works to read, and the examples therein are helpful at illuminating the general thought style, but even with the assistance of the preceding discussion, you will probably find reading this a substantial challenge.

Also on the negative side, *the programs cannot accommodate base models with mere correlations among the exogenous variables*. Most practical models include one or more correlations among the exogenous variables because the researcher does not know the sources of these background covariances, and hence can't avoid them, or because these background correlations are beyond the researcher's current interests. By not allowing correlations among exogenous variables, TETRAD and TETRAD-II demand that the researcher either does the impossible (by specifying the sources of correlations when these are unknown) or the irrelevant (by specifying parts of a model not bearing on the current issue). Should I or anyone be *forced* to account for a correlation between sex and age before we model the effects of these variables on the substantive variables of interest in our model? Clearly not. Yet this is what TETRAD currently demands.

TETRAD-II permits statistical dependencies as being due to single common causes, and hence has moved one step in the direction of permitting covariances, but this is only one step. Covariances may arise from multiple interconnected causes or reciprocal relations, so permitting a single common cause is but a small step towards full incorporation of exogenous variable covariances.[34]

Furthermore, *TETRAD does not permit stacked models or models containing loops or reciprocal effects*. Spirtes, Glymour and Scheines (1993:355-363) seem to have considerable interest in loop models but these as yet have eluded operationalization. Consequently, one could not have found the equivalent models in Chapter 3 using the TETRAD-II program. These models contain loops, and some also contain multiple phantom or indicatorless concepts, which TETRAD-II cannot handle. TETRAD-II also ignores, and hence *does not capitalize on, equality or proportionality constraints*, which might move models from "indeterminate" to clear successes or failures from Spearman's or Thurstone's perspectives.

Glymour et al. caution that the model elaborations suggested by TETRAD "are really intended for the analysis of multiple indicators and may go badly astray for other kinds of initial models" (1987:108). We can surmise that this caution results from some distasteful experiences with particular models. Unfortunately the failure to expose such models to the light of publication day robs researchers of any sense of the styles of model that might be problematic. It is particularly discomforting for those believing age and sex require only single indicators to find that the Purify, Search and MIMBuild modules in TETRAD-II *demand at least two indicators per concept.* Furthermore, no sharing of indicators is permitted.

Also, we must note that if TETRAD or TETRAD-II is used prior to model estimation, the χ^2 for the estimated model is completely untrustworthy.

As a final comment, I might query why TETRAD-II is not being used to shrink models. Tetrad information seems as applicable to suggesting coefficient deletions as insertions, and it seems that only minor program modification would be required to accomplish this.

I believe it is time to integrate TETRAD-II–like procedures and LISREL. This would place TETRAD within LISREL's general notation scheme and would eliminate the need to learn new programming conventions merely to investigate other aspects of one's model. LISREL already contains the information TETRAD-II requires (a data matrix and a model specification), so why should the user have to redo this to run TETRAD?

In the long run, the most fundamental aspect of TETRAD-II may be its use of parallel algorithms (modules) to investigate both structural equation models and Bayesian networks, or categorical causal models. The implications of this rapprochement are far-reaching but they are also beyond the scope of our current purview so we will not pursue this.

4.2.3 TETRAD as Competing with LISREL and EQS

We began Section 4.2 by alluding to Spirtes, Scheines and Glymour's (1990a) and Spirtes, Glymour and Scheine's (1993) observation that when EQS, LISREL and TETRAD-II started from models that were misspecified because one or two necessary coefficients (effects or error covariances) had been omitted, TETRAD-II did substantially better at locating the correct model.

> For a sample size of 2,000, TETRAD II's set included the correct respecification in 95% of the cases. LISREL VI found the right model 18.8% of the time and EQS 13.3%. For a sample size of 200, TETRAD II's set included the correct respecification 52.2% of the time, whereas LISREL VI corrected the misspecification 15.0% of the time, and EQS corrected the misspecification 10.0% of the time.
> (Spirtes, Scheines and Glymour, 1990a:35).

At first glance the superiority of TETRAD II seems insurmountable. At second glance, matters seem even worse. Had Spirtes, Scheines and Glymour (1990a:31) not implemented the test models using an all-η representation,[35] LISREL (but not EQS) would have performed even more poorly because LISREL's standard notation does not permit indicator-to-indicator effects as required by the test models. Hence, without the relatively conscientious modeling by Spirtes, Scheines and Glymour, LISREL would have failed to locate even as many correct models as it did. At third glance, however, the score begins to even out.

The comments by Joreskog and Sorbom (1990) and Bentler and Chou (1990) which followed the Spirtes, Scheines and Glymour paper[36] pointed out that LISREL and EQS were handicapped in the Spirtes, Scheines and Glymour evaluation because they were implemented as beam searches as opposed to breadth searches. That is, LISREL and EQS were programmed to output a single sequence of models. They started from a base model and then added one coefficient at a time, never being able to retract or reconsider any prior coefficients. In contrast, TETRAD examined the full range of first options, and then tried to improve upon each of these firsts with a full range of second guesses, and so on, until all possible combinations of potentially useful additional coefficients were examined. Naturally, one of the key components in TETRAD's search procedure is to try to keep from being swamped by the many possibilities at each step. TETRAD too keeps track of its most promising first step, and the most promising second step that followed the most promising first step, and so on until it has a series of most promising steps, but it does not stop with this series. It reconsiders all the initially less promising model specifications in case an initially modest improvement could be followed by a subsequent superior improvement.

Spirtes, Scheines and Glymour (1990a:37) provided a toned-down version of the above quote after adjusting for the numbers of models the programs provided as suggestions (one guess each for LISREL and EQS, and a variable number of guesses for TETRAD).[37] There was some grumbling that the adjustments were still insufficient, but the basic discussion switched to a concern for equivalent models.

Spirtes, Scheines and Glymour seemed not to have intended to raise the issue of equivalent models, but both Joreskog and Sorbom's, and Bentler and

Chou's reexamination of the models discovered that some of the models located by EQS and LISREL were statistically equivalent to the true target models.[38] The issue this raises is whether locating a statistically equivalent model counts as a success or failure of model search?

This question separates the statistician from the statistically inclined researcher. A researcher's commitment to a substantive domain dictates that he or she view a different, even if statistically equivalent, model as unacceptable. To a researcher in a substantive domain, it makes a difference whether y precedes x (or vice versa), or whether one is confronting an undirected error covariance rather than the statistically equivalent directed effect, because these carry different possibilities for intervention in the real world. The researcher dedicated to a substantive area feels betrayed if a model that is counted as successful includes representations of effects that cannot be trusted to hold in future manipulations of functioning world systems.

In contrast, the statistician feels unjustly put upon if asked to be concerned with issues that are beyond decidability based on the current data. Without a dedication to the substantive domain beyond the current data, the choice between statistically equivalent models seems irrelevant because one has equivalently well modeled the current data, and from the statistician's perspective that is all that counts. For them there is no substantive world beyond, so locating an equivalent model should count as a success.

Joreskog and Sorbom showed their statistician stripes when they argued that "there is no such thing as *the correct model*, neither in empirical work, nor in Monte Carlo studies" (1990:93). Bentler and Chou (1990) showed similar inclinations when they proposed using the "known" population covariance matrix to judge success, as opposed to counting successes based on the true model that created that covariance matrix. In contrast, the philosophers by departmental affiliation (Spirtes, Scheines and Glymour) came across as researcher-oriented when they argued that "not all statistically equivalent models are equally correct" (1990b:109). Hence, on this point there remained substantial disagreement.

The commentators did agree on the artificiality of not allowing substantive knowledge to assist the search. Neither the proponents of LISREL nor of EQS recommend blind sequential addition of parameters.

There was also general agreement that the test was biased in TETRAD's favor because of the relatively small size of the test models. TETRAD's breadth style of searching leads to a geometric increase in the options that have to be searched as a model expands. This leads to a suspicion that, despite its search limiting strategies (Spirtes, Scheines and Glymour 1990a:57), if TETRAD-II confronted models containing many variables, it would tend to bog down, and hence that LISREL and EQS would do comparatively better.

Two further important points seem to have been overlooked in these exchanges. The models used in the Spirtes, Scheines and Glymour tests were biased in TETRAD-II's favor in two crucial respects. First, the true models contained no reciprocal effects. Current TETRAD programming cannot detect reciprocal effects. Hence, any models containing such effects would have provided a 0% success rate for TETRAD, which certainly would have gone a substantial way to evening the final score. Taking a broader view, this program limitation also means that TETRAD is simply not available for anyone wishing to modify a model containing any loops.

Second, all the test models included exogenous variables that were statistically independent. This was hidden by the use of models that had only single exogenous variables, and hence where the need for independence was never raised. Again, had any of the true models contained any covariances between exogenous variables, TETRAD would have suffered more failures because these models are not included within TETRAD's search capabilities. To me, TETRAD's *inability to handle models containing correlated exogenous variables or models with loops are substantial limitations*.[39] These limitations went undetected in the commentary because they were implicitly hidden within the test model's "true" specifications.[40]

4.3 Some Advice and Prospects

The inability to model covarying exogenous variables and the flux in TETRAD's programming lead me to advise researchers, but not statisticians, to temporarily postpone their dive into TETRAD-II. Despite this recommendation, I think there are modifications to our standard operating practices that should be made in response to the issues raised by TETRAD. A basic understanding of the logic underlying the preceding discussions can assist one in doing a more trustworthy job of model development and modification, no matter what program one uses to estimate the model coefficients. Though it may not be obvious, TETRAD-II challenges researchers to become more assertive by examining more user-dictated models and by entering user-directed modifications to models. Several operating principles mesh with this general theme.

4.3.1 Advice

First, TETRAD-II leads me yet again to *strongly admonish researchers to develop competing models prior to estimating any of the models*. Recall the discussion in Chapter 1 that models should not be amalgams of diverse perspectives. Each perspective should be granted its own model. This places a clear emphasis on the substantive issues that separate models or perspectives, and it focuses debate on the boundaries between the relevant theories, and on their points of compatibility and incompatibility. When models are differentially substantively conceptualized, models that are covariance or tetrad equivalent will not necessarily be viewed as theoretically equivalent, and hence covariance or tetrad equivalence will seem simultaneously more and less problematic. It will seem less problematic because it will be clear which of the equivalent models corresponds to the researcher's model, and more problematic because the inability of the current data to distinguish the researcher's theory from an obviously incompatible theory will be substantively discomforting.

Another consequence of creating more separate models will be that the models will, in general, become more parsimonious. More parsimony in turn leads to fewer equivalent models and more assertiveness regarding triad and tetrad constraints among the data covariances. We currently lack a way to reward models specifically for their correct assertion of constraints among the covariances (Spearman's principle), but the penalizing of incorrect assertiveness (Thurstone's principle) is well done via the model χ^2 test and examination of the residuals. Hence, if one locates a sparse model that fits well, the researcher wins on both counts. The sparseness of the model guarantees substantial success in terms of Spearman's criterion and the fit of the model guarantees success in terms of Thurstone's criterion.

More parsimony will also mean more failing models. But again recall from Section 1.1 that there is no reason to demand that all, or even most, published models fit the data in the sense of having an insignificant χ^2. What kind of evidence could be marshaled to demonstrate that most disciplines have been stuck merely one modeling attempt away from the truth, or the true model? It is more likely that we are many steps away, and that each step can only be superseded when the theoretical deficiencies of the prior steps are elucidated, not glossed over. Researchers dedicated to making their model fit, no matter what nonsensical, unconsidered, or theoretically incompatible coefficients they have to insert make it more, not less, difficult to locate the deficiencies of existing models.

Some researchers believe that their place in the Sun, the tabloid, will be jeopardized by any admission of model ill fit.[41] I disagree. Failing models are no less informative and are no less newsworthy than fitting models. I, for one, have no appreciation for scholars whose prominence is purchased by the self-deception that arises from down-playing the fact that a handful of theoretically incompatible coefficients were required to keep the researcher's theory from failing. Science as a community of researchers would seem to function more efficiently without any drag created by consistent unjustifiably-optimistic theoretical assessments.

Another general operating principle is that prior to estimating any model, the researcher should create a mental short-list of the most substantively meaningful and/or likely additional coefficients that might be included *because they match the style of theorizing embodied in the model.* This short-list should be investigated and reported before, or in conjunction with, any data-prompted modifications. Specifically, one should report and discuss whether the selection between nearly equivalent modification indices was arbitrary or non-arbitrary, recalling that equal modification indices often locate equivalent models.[42] Do you see a complementarity between TETRAD-II and Chapter 1? TETRAD-II pushes us to consider breadth searches, Chapter 1 admonishes us to be true to a theoretical perspective and to conscientiously try to verbalize and defend the boundaries between perspectives. That is, Chapter 1 is providing precisely the resource that keeps a breadth search from becoming both mere data dredging, and impossible, because of the multitude of possible options.

If the theorizing in your substantive area is insufficient to draw the boundaries that separate theories or locate the styles of argumentation implicit within theories, then you will have a chance to become your area's premier theorist. All you need do is lay out what you think are the most pronounced fracture lines between the current theories, and the argumentation style implicit in each theory. If you are correct, your demarcations will be a substantive and lasting contribution. If you are incorrect, you will at least be credited for having raised substantively important issues and your vitae will expand because you will be a central part of the next decade of active debates in your discipline.

Implicit in this way of proceeding is a suggestion to start from models that are simple. But do not start from artificially, in the sense of gutting one's theoretical perspective, simplified models.[43] It is the nature of the defense or justification provided for each alternative model that is to be located, respected, and implemented in the model. It is the logic or argumentation style standing behind a model that provides the unity to, and boundaries on, the model. It is disciplinary verbalization fashions that ultimately ground any distinctness to the core of, or boundaries on, a theory. By respecting theoretical boundaries, one is unlikely to do as well at implementing Thurstone's and Spearman's principles as TETRAD does, but fewer glaring mistakes are likely if one spends time examining what it is that creates theoretical-packages within one's discipline. The key objective is to maintain both model parsimony (few coefficients) and theoretical parsimony (few styles of justifications for those coefficients).

Let us push this idea a little further by considering a strategy that should be employed with a well fit model. Imagine we have a well fit model. Imagine also that we can locate a set of effects in this model that are the core of the researcher's theory. Now imagine eliminating those paths from the well fit model and using this pruned model in TETRAD-II. Here we would be using TETRAD-II to seek out models that are equivalent, or nearly equivalent, to the acceptable model, but where it is only the most central package of theoretical claims that is being scrutinized for potential replacement. If TETRAD-II locates even one other equally good model, the central tenets of the original model/theory are still in doubt. Furthermore, if one has difficulty verbalizing or discussing the distinction between the original and nearly equivalent models, one has located a major ambiguity right at the core of the theory. Verbalizing this previously overlooked ambiguity, in your discipline's jargon, will turn you into a first-rate theoretician.

Clearly, only researchers convinced of the theoretical and research utility of equivalent or near equivalent models will appreciate using TETRAD-II with an already acceptable model. And, equally clearly, there is no need to limit this strategy to the TETRAD-II program. There is nothing wrong with eliminating the central components of a theory, and then investigating potentially equivalent replacements for these components, in the context of LISREL.

This raises an issue for the proponents of TETRAD-II. Specifically, when and how should one use substantive knowledge to assist TETRAD-II? Here we are specifically proposing that substantive knowledge of the theory be incorporated to define the boundaries of the search task that is set for TETRAD. This is radically different than specifying a minimal model and then reserving one's substantive knowledge to assist in making the decisions that eliminate some of the numerous possible models that TETRAD proposes. I am inclined not to recommend one strategy over the other. I see both as legitimate depending on the circumstances of the research enterprise. The modular format of TETRAD-II seems to suggest a comparable flexibility but the specific role to be played by theory is not well articulated.[44]

My final piece of general operating advice is oriented toward improving relatively sparse models. *If more than a single modification will be necessary to get a sparse model to fit, pay special attention to seeking coefficients that link together to form omitted indirect effects and/or common causes.* LISREL's modification indices are relatively ineffective at locating omitted chains of coefficients, or pairs of coefficients that are currently missing but required for a variable to act as a common cause (Spirtes, Scheines and Glymour, 1990a:61). You should consciously pursue this possibility to offset LISREL's blind-spots.[45]

4.3.2 Prospects

In much of the above we see tetrad thinking being used to assist one's LISREL analysis, but this is not entirely a one-way street. LISREL can also be used to assist in a tetrad analysis. Bollen and Ting (1993) suggest a way to use LISREL to determine the vanishing tetrads implied by a model. They propose that one arbitrarily specify values for a model's parameters, and use LISREL to calculate the corresponding Σ matrix.[46] One can then calculate all the possible tetrad differences using these model-implied covariances. The tetrad differences within rounding of zero are the model-implied vanishing tetrads. Here one is capitalizing on the fact that LISREL keeps track of the contribution each coefficient makes to each and every covariance during the calculation of the Σ matrix. One then numerically, rather than symbolically, examines whether the covariances imply null tetrad differences. That is, instead of examining a symbolic equation like 4.15 or 4.17, one enters arbitrary values for the coefficients for all the model coefficients, lets LISREL calculate the model-implied covariances, and then uses these covariances in a simple hand calculation of actual numerical magnitude of the tetrad difference of interest. Namely, one uses Eq. 4.9 or 4.16 instead of the equivalent symbolic form Eq. 4.15 or 4.17 to see whether this tetrad difference equals zero or not.[47]

Turning now to slightly more statistical issues, it is a bit too strong to claim, regarding Spearman's principle, that "LISREL ... gives no weight to preserving the ability of the initial model to explain constraints satisfied by the data" (Glymour et al., 1987:120), though it is probably correct to say that current attainment of Spearman's principle is closer to coincidental than planned. A shift toward thinking about the probabilities (likelihoods) that Spearman's and Thurstone's principles are violated or confirmed by any given tetrad (rather than using some cut-point to firmly decide whether there is violation or confirmation), will make it possible to integrate tetrad concerns into maximum likelihood estimation. This move provides likelihoods as a common metric in both LISREL and TETRAD, and should make it possible to make MLE estimates sensitive to the maintenance of progressively more or less strongly implied triad and tetrad differences (Spearman's principle). We may someday see SMLE (Spearman maximum likelihood estimates) that simultaneously minimize the covariance residuals (as currently done by LISREL in the spirit of Thurstone) and the tetrad residuals (the difference between the estimate-implied and observed tetrad differences, in the spirit of Spearman). Strongly implied zero tetrads might be more stringently respected at the time of estimation by forcing some estimates to be precisely zero (eliminating the coefficient from the model) if this results in a valid strongly implied zero tetrad, rather than permitting ordinary MLE to estimate some small and insignificant value, which even though insignificant, still robs the model of the ability to satisfy Spearman's criterion. And we may even see Spearman modification indices. That is, modification indices that carry a built-in penalty for the loss of strongly implied null tetrads (Spearman's principle) which could counteract the current magnitude of pure χ^2 gain (Thurstone's principle), which drives the current modification indices.[48]

On another statistical front, the enterprise to date is missing an important statistical demonstration, namely that a model that is well fit by the usual estimation methods can still be *substantially improved upon* by paying attention to tetrads. Without such a demonstration one has difficulty responding to researchers who claim not to need TETRAD because their models fit well. If TETRAD cannot improve upon a model that is acceptable by standard criteria (χ^2, residuals), then all we need do is satisfy those criteria and forget about TETRAD because its criteria will be of no further help. Admittedly, it could potentially provide equivalent alternatives, but the justification for paying attention to alternatives is very different from a justification for potential direct improvement of an already acceptable model.

And we may also have to rethink another statistical cornerstone. Bollen and Ting (1993) suggest that we imagine confirmatory tetrad analyses (CTA) in contrast to exploratory tetrad analyses (ETA). That is, instead of using tetrad differences to locate the true model, use the attainment of implied tetrad differences to test predetermined models. Bollen and Ting (1993) suggest that tetrad-nested models (models where the vanishing tetrads for one model are a subset of those implied by another model) may lead to the ability to compare (via the nested tetrads) models that are not nested in terms of their coefficients and basic likelihood ratio χ^2.[49] And, since the tetrad procedure does not require parameter estimates, this raises the possibility that some underidentified models might be assessed, against other identified or underidentified models, via their tetrad implications. If a currently underidentified model proved to function comparatively well tetrad-wise, one would have additional incentive for identifying the model via the collection of additional data constraints arising from judiciously placed new variables (Bollen and Ting, 1993). The degree of nonredundancy and the utility of the tetrad χ^2 versus the traditional likelihood ratio χ^2 in real research contexts remain to be demonstrated.

In sum, TETRAD is fostering a major shift in emphasis from using statistics to estimate the coefficients in models, to using statistics to assist in locating the proper model in which to estimate coefficients. In the future, researchers will have to be concerned with gathering data that can distinguish between statistically equivalent models, and they will also have to be cognizant of the fact that particular substantive theories may be consistent with some, but not other, statistically equivalent models (Spirtes, Scheines and Glymour, 1990b:109-110).

Notes

1. Contrast the discussions in Chapters 1 and 2 with TETRAD-II's requirements for at least two indicators in the Purify subroutine, and the mandatory use of unidimensional or purified indicators in the multiple indicator model building subroutine MIMbuild (Scheines et al., 1994).

2. Other sets of rules for locating equivalent models can be found in Lee and Hershberger (1990) and Lee (1987).

On another statistical front, Cudeck and Browne (1992) discuss the estimation of models that do not hold exactly in the population. In this context, a refined definition of equivalence will be required because there is no current consensus on whether equivalence should refer to the overall fit matrix, or to the exact and perturbation matrices into which the overall matrix is decomposed, or whether equivalence should refer to the exact matrix alone. I would be inclined to refer to the equivalence of the overall matrix because model claims about what constitutes an error or perturbation is part of the model's claims. If the equivalence of the exact matrix alone is accepted as sufficient for model equivalence, the number of equivalent models will increase substantially because a wider set of models might fit only approximately.

3. See Section 4.2.1.3 for the basic equation for partial correlations, and see Blalock (1964, 1979) or Duncan (1975) for the historical context of the connection between partial correlations and standardized regression slopes in recursive models.

4. All these models are saturated with effects. They imply a Σ that is identical to the data matrix S, they have zero residuals, and a χ^2 of zero with zero degrees of freedom.

5. If there are several correlated exogenous variables, these are given an arbitrary ordering and the following rules are applied, disregarding the possible reorderings of the exogenous variables among themselves (see Stelzl, 1986:313).

6. The argument here is nearly the converse of the observation that adding a control variable changes a partial standardized slope, or the corresponding partial correlation. If the original partial correlation and slope linking X_3 to X_5 is zero, a new partial correlation with an additional control variable will in general be non-zero.

7. Partial correlations are calculated from the basic correlations among the variables (Section 4.2.1.3), so any one set of observed correlations cannot be consistent with both of these partial correlations being zero.

8. If two or more null effects are introduced, Stelzl (1986:318) recommends applying Rule 1 to select out the tentatively equivalent models for each of the null effects individually, and then selecting as truly equivalent only those models appearing on all the lists of tentatively equivalent models. Obviously this will become a tedious and error-prone procedure for anything but the smallest of models.

Lee and Hershberger (1990:318-319) provide a slightly different way of demonstrating the equivalence of models that unifies Stelzl's rules, but the direct connections to partial correlations are lost in their representation. The Lee and Hershberger approach may be slightly simpler but it is less useful here because we will be depending on the connection to partial correlations when we draw parallels to TETRAD later in this chapter.

9. See *Essentials* (145, 154, 162), but do not create artificially sparse models by risking the elimination of likely effects. This can lead to biased estimates (*Essentials*, 147) and can render the diagnostics proposing specific model revisions less effective.

10. Recall *Essentials* (171-173).

11. Recall *Essentials* (177).

12. The equations are easily reproduced by starting from the correlation pair with the four variables in sequential order, and then switching the inner two subscripts to get the correlation pair on the right-hand side of the first equation. Using the outer pair then inner pair of original subscripts provides the correlation pair on the right side of the second equation.

13. *Essentials*, Eq. 1.54: $r_{xy} = \mathrm{Cov}(x,y)/\sigma_x\sigma_y$.

14. Recall *Essentials* (26-31, 106-117).

15. Labeling the bottom line of variables y_1 to y_4 would make this appear more like a factor model, but would provide notational inconsistencies when we include the types of effects depicted in the lower portions of this figure.

16. Recall Eq. 1.1 above, or see *Essentials*, Eq. 1.67, or any off-diagonal element of *Essentials*, Eq. 4.37, with a single η.

17. If you feel uncomfortable about this, use Duncan's procedure (Section 1.3.3) on these covariances.

18. We begin with the four equations comprising the middle model in Figure 4.2:

$$\eta_1 = \beta_{1,5}\eta_5 + \zeta_1$$

$$\eta_2 = \beta_{2,5}\eta_5 + \beta_{2,1}\eta_1 + \zeta_2$$

$$\eta_3 = \beta_{3,5}\eta_5 + \zeta_3$$

$$\eta_4 = \beta_{4,5}\eta_5 + \zeta_4.$$

For the $\mathrm{Cov}(\eta_1,\eta_2)$, we select the equation for η_2 and multiply by η_1, using the names for the variables on the left and the model's claims about the sources of these variables on the right:

$$\eta_1\eta_2 = (\beta_{1,5}\eta_5 + \zeta_1)(\beta_{2,5}\eta_5 + \beta_{2,1}\eta_1 + \zeta_2).$$

Multiplying out the parts gives

$$\eta_1\eta_2 = \beta_{1,5}\beta_{2,5}\eta_5\eta_5 + \beta_{1,5}\beta_{2,1}\eta_1\eta_5 + \beta_{1,5}\zeta_2\eta_5 + \beta_{2,5}\eta_5\zeta_1 + \beta_{2,1}\eta_1\zeta_1 + \zeta_2\zeta_1.$$

Taking expectations of both sides of the equation, remembering that structural constants can be factored outside the summation implicitly involved in taking expectations (*Essentials*, 4, 339), provides

$$\mathrm{Cov}(\eta_1,\eta_2) = \beta_{1,5}\beta_{2,5}E(\eta_5\eta_5) + \beta_{1,5}\beta_{2,1}E(\eta_1\eta_5) + \beta_{1,5}E(\zeta_2\eta_5)$$
$$+ \beta_{2,5}E(\eta_5\zeta_1) + \beta_{2,1}E(\eta_1\zeta_1) + E(\zeta_2\zeta_1).$$

The expected value of the product of η_5 and any error variable is the covariance between η_5 and that error, which we assume to be zero. And the covariances between the errors are also assumed to be zero:

$$\mathrm{Cov}\ (\eta_1,\eta_2) = \beta_{1,5}\beta_{2,5}\mathrm{Var}(\eta_5) + \beta_{1,5}\beta_{2,1}\mathrm{Cov}(\eta_1,\eta_5) + \beta_{2,1}\mathrm{Cov}(\eta_1\zeta_1).$$

Neither $\mathrm{Cov}(\eta_1,\eta_5)$ nor $\mathrm{Cov}(\eta_1\zeta_1)$ is a fundamental model parameter, and so these must be expressed as functions of the basic model coefficients. Taking the equation for η_1, multiplying by η_5, and following Duncan's steps provides

$$\eta_1\eta_5 = \beta_{1,5}\eta_5\eta_5 + \zeta_1\eta_5$$

$$\mathrm{Cov}(\eta_1,\eta_5) = \beta_{1,5}\mathrm{Var}(\eta_5) + 0.$$

Similarly, using the equation for η_1, multiplying by ζ_1, and following Duncan's procedure provides

$$\eta_1\zeta_1 = \beta_{1,5}\eta_5\zeta_1 + \zeta_1\zeta_1$$

$$\mathrm{Cov}(\eta_1\zeta_1) = 0 + \mathrm{Var}(\zeta_1).$$

Substituting these into the equation for $\mathrm{Cov}(\eta_1,\eta_2)$ expresses this covariance as a function of the basic model coefficients:

$$\mathrm{Cov}\ (\eta_1,\eta_2) = \beta_{1,5}\beta_{2,5}\mathrm{Var}(\eta_5) + (\beta_{1,5})^2\beta_{2,1}\mathrm{Var}(\eta_5) + \beta_{2,1}\mathrm{Var}(\zeta_1).$$

Selecting the equation for η_2, multiplying by η_4, taking expectations and similarly reducing an emergent $\mathrm{Cov}(\eta_1,\eta_5)$ results in

$$\mathrm{Cov}(\eta_2,\eta_4) = \beta_{2,5}\beta_{4,5}\mathrm{Var}(\eta_5) + \beta_{2,1}\beta_{1,5}\beta_{4,5}\mathrm{Var}(\eta_5).$$

We now have the desired covariances expressed as implications of the model's structural coefficients.

19. This is not identical to the equation provided by Spirtes, Scheines and Glymour (1990a:11), which I think contains an error, but it has the same substantive implication.

20. The "falsification" and "explanatory" principles are the titles used in Spirtes, Scheines and Glymour (1990a:13) while the Thurstone and Spearman titles are used in Glymour et al. (1987).

21. Our top model has one fewer coefficient than the middle model, though we purposefully attempted to render this point moot by the $\eta_{\#}$ "allusion to potentially numerous, as yet unspecified, model parts" where either model might have relatively many or few coefficients.

22. *Essentials* (46-49).

23. Higher order partial correlations are discussed by Blalock (1979:461). Glymour and Spirtes (1988:184) and Glymour et al. (1987:75-83) discuss the connection between partial correlations and tetrad differences.

24. Consider this first order partial correlation as paralleling Stelzl's discussion of a model in which only one common cause brackets the two variables with the mandatory null effect. Higher order partial correlations are required if there are more common causes.

25. You can demonstrate this by using the variables η_1, η_2 and η_5; obtaining the covariance between η_1 and η_2 from note 18; calculating the covariances between η_5 and both η_1 and η_2 using the procedure demonstrated in note 18; and substituting these values into the following rearrangement of the equation that appears just above this paragraph. We rearrange this equation so it is expressed as a *possible* equality, and insert the definition of each correlation as a covariance divided by the corresponding variables' standard deviations:

$$\frac{\mathrm{Cov}(\eta_1,\eta_2)}{\sigma_1\ \sigma_2} = \frac{\mathrm{Cov}(\eta_1,\eta_5)}{\sigma_1\ \sigma_5}\ \frac{\mathrm{Cov}(\eta_2,\eta_5)}{\sigma_2\ \sigma_5}.$$

Multiplying both sides by $\sigma_1\sigma_2\sigma_5\sigma_5$ provides

$$\mathrm{Cov}(\eta_1,\eta_2)(\sigma_5)^2 = \mathrm{Cov}(\eta_1,\eta_5)\mathrm{Cov}(\eta_2,\eta_5).$$

The variance of η_5 is a nondecomposable model coefficient. When you substitute the covariances from above into this equation, you will find that this equality does not hold unless the coefficient β_{21} is zero for the middle or bottom models, though it holds unconditionally if the equations corresponding to the top model in Figure 4.2 are used.

26. One detail is the appropriate level of significance to use. The usual .05 is very restrictive. A larger value is recommended, but see note 30 below.

27. Glymour et al. (1987) discuss the details of how this was attempted, though their discussion is probably too detailed to interest anyone with less than a few days to spare.

28. This is primarily accomplished by starting from a very simple base model and then adding as few coefficients (effects) as possible. This leaves the Spearman and Thurstone principles as the main driving forces behind most TETRAD-induced revisions.
29. For example by making the program sensitive to logically impossible conditions. It is impossible to move from a model that does not currently strongly imply a particular zero tetrad difference to one that does by adding model coefficients.
30. TETRAD could be programmed to avoid this particular decision by calculating the actual probability of each difference and using that probability as a weight in the decision process, rather than attempting to defend a particular probability level as the proper all-or-none decision criterion.
31. See also Glymour, Spirtes and Scheines (1991a:176) and Verma and Pearl (1990).
32. Some edges may remain undirected and TETRAD-II proceeds to investigate these as it pursues its final set of suggested model specifications.
33. For example, imagine two variables with no direct effects between them but depending on two common causes, where only one of the common causes is included in the current data model and data set. Theorem 1 will suggest an "edge" is required because the variables are not independent when one of the two common causes are controlled, but Theorem 2's attempts at "directing" the edge should be useless as there is indeed no direct effect between the variables to directionally orient. See also Spirtes, Glymour and Scheines (1991a:30).
34. I suspect that the programming used to permit correlations among the error variables may be adaptable to allowing correlations among the exogenous variables. If one is starting from a null model, the tricky task is to find a principled way to decide where to permit mere correlations when the program has nothing but correlations as input data. TETRAD-II might insert merely correlated variables where it has difficulty with the second step of directing several edges for a variable, but the issue remains as to how in principle to characterize how much, or what style of, difficulty is warrant for insertion of a mere correlation. By inputting a basic model with specified locations for mere correlations into TETRAD the user could provide an invaluable resource. I see no problem in principle with allowing covariances (see Duncan, 1975:36-41 for the implications of this) but this will demand some fundamental changes to the basic core of routines and possibly development of some new decision criteria.
35. *Essentials* (209-212).
36. Bollen (1990b) provided an additional comment that is likely to lead to substantial improvements in TETRAD-II. Bollen provided the mathematics illustrating how the influence of outliers on tetrad differences can be detected (though a justification for the routine elimination of influential outliers is still lacking). He also provided a significance test of tetrad differences that avoids the usual multivariate normal assumption, and polished TETRAD's approach to handling the significance of multiple tetrad difference tests. These modifications are likely to improve TETRAD's performance, and hence are likely to be implemented without delay, but they are not central to the issues of concern to us here.
37. "Were TETRAD-II to be as bold as LISREL VI or EQS, its single model reliability at sample size 2,000 would drop from 95% to about 42.3%. ... At sample size 200, TETRAD-II's single model reliability was 30.2%" (Spirtes, Scheines and Glymour, 1990a:37).
38. These were located by spotting equal, or nearly equal, values in the modification indices. The models included the statistical equivalence of additional effects between indicators and error covariances, and the equivalence of a structural model containing a loop and a recursive model.

39. Advocates of factor analysis and the two-step procedure will see this as a substantial limitation because it means TETRAD is unavailable if one's model has two or more correlated factors, or if one is trying to assess the step-one measurement model of the two-step procedure.

40. In working through Chapter 10 of Glymour et al. (1987), I thought I had discovered an error in the proof of Lemma 11 on page 279, where it seemed that several exogenous (nominally independent) variables (denoted by set I) were described as being necessarily statistically independent of (and hence uncorrelated with) one another. A call to Clark Glymour (December, 1991) clarified that set I was supposed to be a set of statistically independent variables, and not merely exogenous (nominally independent) variables.

Being a nitpicking academic I went back to discover where I had gone wrong, only to find more than a minimal amount of ambiguity in the definitions of "independent variable" (1987:261), in the set of independent variables (I) (1987:263) and in independent equations (1987:263). The example following their definition 9 (1987:263) ends with the phrase "because v is not an independent variable." Since statistical independence is a relation between two variables, it seemed clear that the use of "independent variable" here, and in the definition this is an example of, must refer to independence in the sense of their definition 6 (page 261) where this is any variable that does not have directed effects pointing at it (the usual statement of an exogenous variable). But this is wrong. I will let Glymour et al. explain why, because I can't!

Glymour et al. can make a few minor changes that will make their definitions consistent with their proofs, but to do so will force them to come clean about their unavoidable insistence on the absence of exogenous variable covariances in any model to be entered into TETRAD or TETRAD-II. This is not an academic triviality, it is a substantial limitation.

41. These are often the same researchers who view passing the .05 probability level for χ^2 as the crucial test of fit. Recall the discussion of more appropriate χ^2 probabilities in Section 2.3.

42. Freeing coefficients with equal modification indices are likely to result in at least narrowly equivalent models, but these may or may not be broadly equivalent models (Luijben, 1991:658).

43. One of the reasons for the relatively poor showing of EQS and LISREL was that when these programs began with an incorrect start model, the model misspecification led to inaccurate parameter estimates (or a failure to converge) which subsequently had detrimental effects on the utility of the information provided by the modification indices. This exemplifies why backward elimination of excess (unnecessary) coefficients may be preferred to forward selection.

44. If the TETRAD-II proponents believe that in many cases "the necessity of theory is badly exaggerated" (Spirtes, Glymour and Scheines, 1993:133) it seems incumbent upon them to articulate when, if ever, theory continues to be necessary, and what it is necessary for, in the context of TETRAD-II. If theorizing merely consists of melding observed time sequence to common sense (e.g., Spirtes, Glymour and Scheines, 1993:138), then theory may indeed be of little use. Loops and reciprocal causes eliminate the necessity and utility of time sequencing and the commonness of sense-making often robs it of consistency and principles.

45. This meshes well with Bentler and Chou's (1990) suggestion to slightly overfit one's model, and then to cut back on the excess coefficients before adding other coefficients.

46. It would be permissible, and even advisable, to use the maximum likelihood estimates but these are not mandatory. Even incorrect estimates should preserve the vanishing tetrad implications since these do not change if the magnitudes of the coefficients change. The vanishing tetrads should hold no matter what coefficient values are used. The ability to enter arbitrary values makes the procedure applicable to underidentified models.

47. Triad differences, or partial correlations, can be similarly examined.

48. Given the emphasis placed on the Spearman and Thurstone principles in TETRAD, the near absence of these from TETRAD-II is remarkable. The current draft of the TETRAD-II user's manual does not list these in the index, or discuss reasonable ways of doing the weighting that implicitly connects the Spearman and Thurstone principles in the Purify and Tetrad modules.

49. Bollen and Ting (1993) also provide improved statistical procedures of making an overall model χ^2 test of the model-implied vanishing tetrads, which presumably is not redundant with the usual χ^2 test.

Chapter 5

Stacked Models with Differing Sets of Indicator Variables

Imagine we have two groups on which we have gathered comparable data. We could create a covariance matrix for each of these groups and estimate separate, though similar, models for the two groups. Or we could use both the data matrices and groups' models to create a stacked, or multisample, model in which a single LISREL command file and a single set of maximization iterations produces estimates for the two groups simultaneously.[1]

Stacked models have been used relatively infrequently even though there is nothing difficult about specifying or estimating such models.[2] This may reflect a failure to appreciate what can be gained from stacked models. The primary use of stacked models is to locate conditional differential effectiveness, or what is technically known as interaction. Conditional effectiveness, in turn, is a fountain of practical utility. If an effect leading to the target variable is stronger under one condition than another, then one can most effectively intervene by guaranteeing that the relevant condition holds, or preserve the current value of that variable by guaranteeing that the condition does not hold.

Stacked models can also assist by localizing any differential effectiveness. A stacked model can be used to test whether the only differences between the groups are in the measurement portion of the model or whether the structural models differ despite equivalent measurement coefficients.[3] But the testing for between-group equivalence of the measurement and structural models should not be thought of as rigidly prescribed alternatives. These should be viewed as possible, rather than mandatory, or even recommended, tests. Recall that Chapters 1 and 2 argued against the routine separation of the measurement and structural models. In this spirit, it would seem more appropriate to tailor the specification of the groups to correspond to potentially manipulable variables and to select between-group constraints matching the researcher's substantive concerns. The fact that the required constraints fall in the measurement or structural segments of the models are of little inherent interest.

A further advantage provided by stacked models is that they permit one to estimate intercepts in the conceptual level equations, and to test for mean differences between the groups. The modeling here is a bit more complex, but it permits one to predict specific expected values of the dependent concepts for individuals possessing specified combinations of values on the independent concepts, and to test for potential biases in supposedly equivalent indicators. The relevant procedures have been discussed in several sources (e.g., Joreskog and Sorbom, 1988:229-259; *Essentials*, 276-322), so we need not repeat these discussions here. Instead, this chapter attempts to describe other useful things that stacking can accomplish which are relatively, or entirely, unknown.

All the uses considered below demand the stacking of models that are based on different numbers of observed variables. The LISREL 7 manual devotes a chapter to multiple sample analyses, but all of the examples in that chapter have the same number of indicator variables in the stacked groups. And, if one tries to enter a stacked model with differing numbers of indicators in LISREL 6 or 7, one receives any of a variety of fatal error messages depending on the specification of the model.[4]

Section 5.1 delineates the additional research opportunities arising from stacking models based on different numbers of indicator variables. These opportunities include identifying normally unidentified models, controlling for unmeasured variables, testing for the sufficiency of known mechanisms, and integrating nonparallel model segments and data sets. Section 5.2 presents an extended example illustrating one of these uses and simultaneously introduces the bit of modeling trickery that permits one to circumvent LISREL's requirement of equal numbers of indicators in all the groups. I have attempted to minimally, though completely, discuss the modeling trickery in anticipation that LISREL will be reprogrammed to permit different numbers of indicators in multiple groups. Once this is done, the modeling contortions will become unnecessary, and the focus of this chapter will rightfully return to the research opportunities provided by the stacking of models based on differing sets of indicator variables.

5.1 The Utility of Stacked Models with Unequal Numbers of Indicators

For the models discussed in this section, we imagine that the model segment depicting one of the stacked groups contains an indicator variable that is unavailable in the other group. This may be an extra characteristic or question asked of the employed but not the unemployed, the delinquents but not the nondelinquents, the experimental subjects but not the controls, the blacks but not the whites, or the females but not the males. If there are several additional indicator variables available in one of the groups, these can be utilized in a parallel fashion.

5.1.1 Identifying Underidentified Models

Figure 5.1 illustrates several models where one of two groups in a stacked model is underidentified if this group is modeled separately, but where the stacked model including this group is identified if estimated with the indicated constraints between the groups. In the following, the underidentified group usually appears as the lower of the two groups in the figure, and the second of the groups in the command file, but this placement is purely arbitrary. The broken curved lines in Figure 5.1 indicate equality constraints between the coefficients to which they are attached in the two groups. The substantive research gain made by stacking in these instances is that an otherwise underidentified model segment, usually the reciprocal relationship in the lower group, is now identified and estimable. These models support the claim that *stacking models with unequal numbers of indicators may provide a means of estimating complex causal structures that have defied other attempts to attain model identification.*[5]

The key to these models is the use of equality constraints between the groups to strengthen the model's constraints in the second group, and hence to limit the range of possible coefficient estimates that simultaneously satisfy the data and model constraints in that group. The reciprocal effects in group 1 at the top of Figure 5.1A would be uniquely identified, and hence could be estimated if this group were run as a nonstacked model.[6] Group 2 at the bottom of Figure 5.1A, on the other hand, is underidentified if one attempts to estimate it alone because it lacks a variable that uniquely causes the variable at either, but not both, ends of the reciprocal relationship, as in group 1. Group 2 also lacks the indicator(s) of this unique conceptual cause, and this creates the need for a procedure permitting the entry of differing numbers of indicators in stacked models. With one of the group 2 reciprocal effects specified, through its equivalence to the corresponding and identified effect in group 1, the reverse effect will be identified by the covariance between the variables at the ends of the reciprocal effects in group 2.

Since only one additional constraint is required to identify the reciprocal relation in group 2, the researcher has a choice as to whether to constrain the right-pointing or left-pointing effect (compare Figures 5.1A and 5.1B). Only the

Figure 5.1 Stacking for Identification.

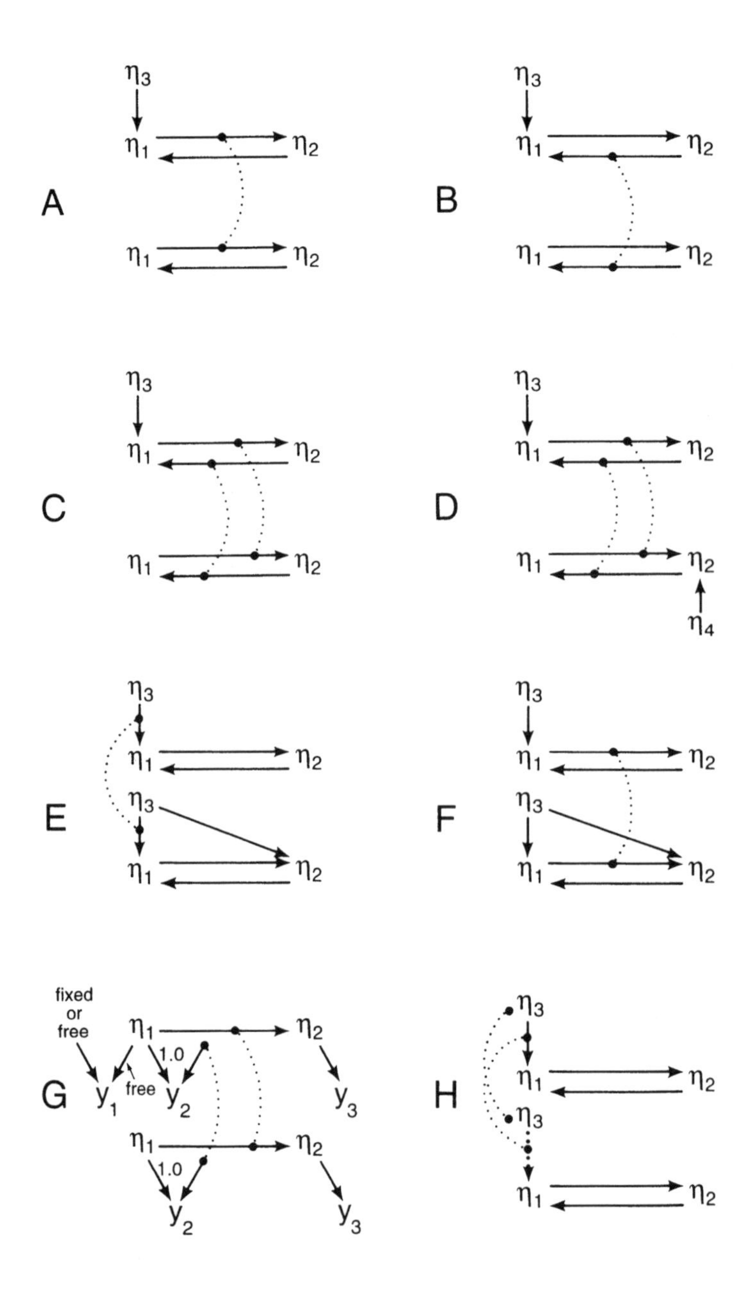

effect whose equality between the groups is most substantively reasonable, and hence least contentious, should be constrained to be equal across the groups. It should be obvious that whichever of these constraints is used, this amounts to simply inserting the estimate from the identified top group into the model segment for the bottom group. That is, since the reciprocal effects in the bottom group are underidentified, this group has no leverage with which to alter the corresponding estimate in the top group to which the coefficient is constrained.[7]

If both the left- and right-pointing effects are constrained, as in Figure 5.1C, the lower segment of the stacked model will be overidentified, and there will be an extra degree of freedom available to do a test. For example, using the χ^2-difference procedure to compare a stacked model with just the left-pointing effects equal between the groups to a model with both the effects constrained could be described as a test of the equivalence of the right-pointing effects.[8] That is, not only can a stacked unequal-indicator model lead to estimates of the entirely reciprocal model segment for group 2, it can in fact lead to a test of whether applying both the between-group constraints simultaneously is consistent with the data.[9]

The model in Figure 5.1D is even more strongly overidentified than the model in Figure 5.1C. The reciprocal effects in each of the model segments individually are identified, so the between-group constraints on the reciprocal effects contribute two extra degrees of freedom. The substantive point at issue in this model is not identification, but whether the estimates arising from using a unique cause of one of the reciprocally related variables is consistent with the estimates arising from attaching an entirely different unique variable to the other of the reciprocally related variables. Obviously, if η_3 in the top group had more or fewer indicators than η_4 in the bottom group we would require a procedure for stacking models with unequal numbers of input variables, but here the procedure leads to substantial overidentification, not just identification.

In fact, the diagnostics accompanying this model might point to model features other than the current reciprocal effects as being problematic. For example, an effect from η_3 to η_2 in the top group, or a reciprocal effect from η_1 to η_3 in the top group, would now be identified and might be called for more strongly than relaxation of the equality constraints on the current reciprocal effects. These effects would have been underidentified had these effects been added to the top group individually, but these would now be identified because of the overidentifying restrictions on the existing reciprocal effects. Hence, Figure 5.1D points toward a whole new range of diagnostic possibilities which arise because additional model coefficients are identified and these provide new options for model modification.

Figures 5.1E and 5.1F extend the range of models to which the use of between-group constraints to identify models is appropriate. Here the model segment for the lower group is underidentified, not because the concept η_3 is missing, but because η_3 is thought to have an additional effect, namely an effect on η_2. Without this effect the bottom model segment is identified, but with this effect the bottom model segment is underidentified. That is, the second group seems to have too many effects, not too few variables. The stacked model segments in Figure 5.1E and 5.1F may or may not be based on the same set of indicators, depending on whether or not any one of the concepts is missing an indicator in either of the groups.

Either of the constraints illustrated in Figure 5.1E or 5.1F will identify the lower model segment but constraining one of the reciprocal effects would be preferable if this equality is in fact substantively reasonable. A constraint on the effect of η_3 on η_1 only indirectly assists in the identification of the reciprocal effects. This will result in larger standard errors for the reciprocal effects than if one constrains either of the reciprocal effects directly. But this gain in efficiency would be insufficient to justify selection of model 5.1F over 5.1E unless the substantive claim about the equality in Figure 5.1F is at least as reasonable as the constraint in Figure 5.1E.

The model in Figure 5.1G changes the issue from identification of reciprocal effects to identification in the measurement part of a model. It is not uncommon to find that a concept η_1 can be measured in two ways, one expensive, the other cheap. In Figure 5.1G the preferable and expensive way of measuring η_1 provides the indicator y_1. The less expensive way of measuring η_1 provides questionable measurement quality, and hence there is a need for estimating the quality of the cheaper y_2 measurements. The researcher scrapes together sufficient funds to use the expensive y_1 measure with some of the subjects (those in the top or first group), but cannot afford to use it on the rest of the subjects (group 2). Given that the only measure of η_1 available in group 2 is the measure in which the researcher admittedly has little confidence, the use of the fixed $\Theta_{\epsilon 2}$, as discussed in Chapter 1, is questionable. Hence, the $\Theta_{\epsilon 2}$ should be estimated, but no models containing a free estimate for $\Theta_{\epsilon 2}$ could be run for group 2 alone because, with a single indicator, this would make this part of the model underidentified.

The researcher could estimate the model for group 1 and then use the resulting $\Theta_{\epsilon 2}$ value as a fixed coefficient in a separate run for group 2, but this would mean the researcher would be relying upon, and hence would have to defend, the idea that all the information about the quality of the questionable y_2 indicator is provided by the first group. The researcher would prefer to get an estimate of $\Theta_{\epsilon 2}$ that is a compromise estimate that acknowledges that both groups provide information on the quality of y_2 as an indicator, especially if there are many more cases in the second group.

A stacked model with unequal numbers of indicators (Figure 5.1G) is again the answer to the researcher's problem. Constraining the $\Theta_{\epsilon 2}$'s to be equal between the groups *assists* in making a compromise estimate that permits both the group 1 and group 2 measurements of y_2 to contribute to the assessment of

y_2's measurement error variance. But the equality constraint on $\Theta_{\epsilon 2}$ is merely an assistance in doing this, it by itself is insufficient to do this. If the equality constraint on $\Theta_{\epsilon 2}$ was the only equality constraint, all this would do is force the estimate arising from group 1 onto group 2. Because the $\Theta_{\epsilon 2}$ coefficient is underidentified in group 2, group 2 has no leverage to change whatever estimate is foisted upon it by group 1. You will recognize this as precisely the same problem the researcher would have had if two separate models were run, and the estimate from group 1 was applied to group 2.

The resolution to the researcher's problem arises from noting that η_1 is supposed to be the same concept in the two groups, and hence that it should display the same effects in the two groups. If we make use of this fact by also forcing β_{21} to be equal in the two groups, then $\Theta_{\epsilon 2}$ becomes overidentified, not just merely identified in the group 2 portion of the stacked model, and group 2 regains some leverage which permits it to contribute to a compromise assessment of y_2's measurement quality. That is, the covariances in group 2 can now move the estimate of $\Theta_{\epsilon 2}$ away from the estimate that would be found in group 1 alone, and hence both groups contribute to the assessment of the quality of the y_2 measurements. All of η_1's incoming and outgoing effects should be constrained to be equal in the two groups, since these effects presumably should not differ merely because one was able to measure y_1 for some of the sampled cases, namely the group 1 cases. The more such constraints, the more strongly overidentified is $\Theta_{\epsilon 2}$, and the more leverage group 2 will have.

Note again, that to implement such constraints in a stacked model, we must be able to stack models based on differing numbers of indicator variables. Without this, our researcher is unable to arrive at a compromise assessment of the measurement quality of y_2 to which the second group contributes.

The $\Theta_{\epsilon 1}$ for the gold standard indicator y_1 might have been fixed or free in the preceding discussion. Given the discussion in Chapter 1, you will recognize that if $\Theta_{\epsilon 1}$ is left free the researcher is admitting some degree of uncertainty in the conceptualization of η_1. Note also that in Figure 5.1G y_2 is used as the indicator that scales the concept η_1, but this is not necessary. One could use y_1 to scale η_1 in group 1, and equate λ_{21} in the two groups to provide η_1 a comparable scaling in group 2. One would then use this estimated lambda value as a fixed coefficient in place of 1.0 in future models to maintain η_1's scale as reflecting that of y_1 despite subsequent use of only the y_2 indicator.

Figure 5.1H attempts to identify the reciprocal effects in group 2 by using a stacked model based on differing sets of indicators, but it tries to do this by setting it up in a different way than above. This model has been left until the end because this attempt at identification is unsuccessful, though the reason for the failure is worth considering. Figure 5.1H proposes that instead of directly constraining one of the identified reciprocal effects in group 1 to equal the corresponding effect in group 2, we attempt to identify the relation by constructing a group 2 variable that, hopefully, will do the identification in group 2 just as it does in group 1. Namely, we try to provide group 2 a unique cause for η_1 that behaves like the η_3 in group 1. The advantage of this is that if this were to work it would circumvent the need for any equality constraints on the reciprocal effects between the groups, and hence we could get separate and untainted estimates of the reciprocal effects in both the groups.

In fact, the model in Figure 5.1H is not identified. To see why, consider the reason unique causes of the variables at the ends of reciprocal relations are usually able to identify the reciprocal effects. The presence of a unique causal variable focuses an additional data constraint on the reciprocal effects. In particular, the covariance between the unique cause and the variable at the far end of the reciprocal relation breaks the symmetry of the reciprocal effects. The effect of the unique cause on the far-end η uses one of the reciprocal effects more heavily than the other. Whichever of the reciprocal effects is used first in moving from the unique cause to the far-end η is most heavily implicated because it provides the basic indirect effect that is enhanced by subsequent, and necessarily weaker, cyclings of effects through the reciprocal loop.[10]

The implicit problem with the Figure 5.1H model is that there is no rigid data covariance to constrain the covariance between the manufactured η_3 and η_2 in group 2. The covariance between these η's, like all η covariances, is an artifact implied by the functioning of the effects in the model.[11] Any set of reciprocal effects whatever, in conjunction with the β_{13} imported from group 1, will subsequently imply some particular covariance between η_3 and η_2, but given that there is no constraint whatever on this covariance (due to the lack of η_3 indicators in group 2) this implied covariance is not held in check by anything in the model. Hence this covariance is free to vary and lacks the constraining ability that normally assists in breaking the symmetry of reciprocal effects. In fact, when this model is estimated, several classic signs of identification problems arise, namely warnings of possibly unidentified coefficients, and an inability to calculate standard errors, T values, or modification indices.[12]

5.1.2 Controlling for Unmeasured Variables

Figure 5.2 illustrates a different reason for stacking models that are based on differing numbers of indicators. In these models both the groups are individually identified, so identification is not at issue. What is at issue is the ability to use specific variables as control variables, and hence the ability to gain effect estimates that can be interpreted as the effects arising even after specific other variables have been controlled.

Figure 5.2A corresponds to a situation in which all of the causes of η_1

Figure 5.2 Stacking for Control of Unmeasured Variables.

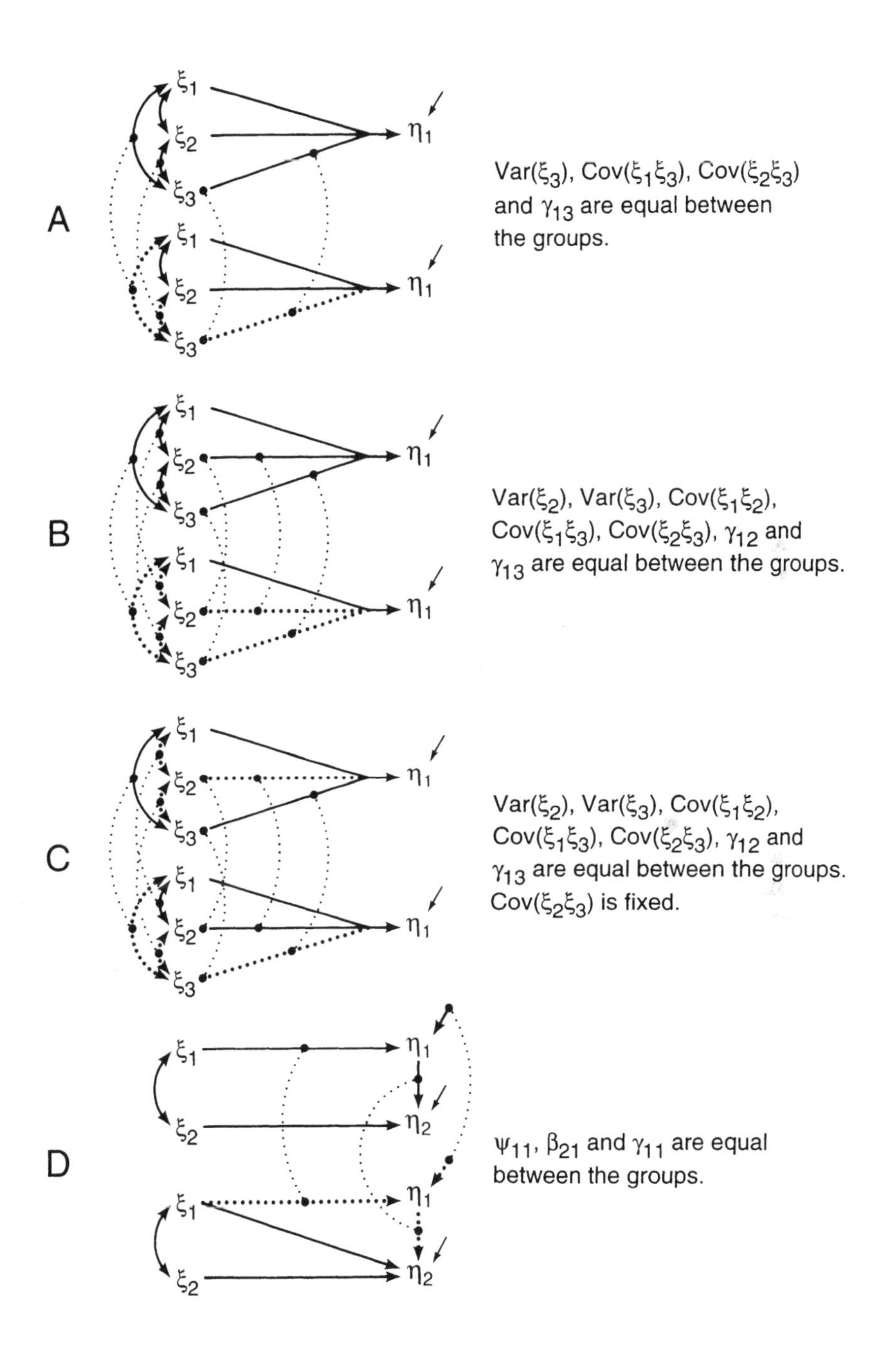

were measured in the first group, but the indicators of concept ξ_3 are missing from the second group. It makes no difference whether the indicators of ξ_3 are missing due to oversight, mistake, machine failure, the respondents' refusal to answer, or whether the indicators of ξ_3 were too expensive to gather for all the subjects.

Given the absence of ξ_3's indicators from the second group, the researcher has a choice. Either do no controlling for ξ_3 in group 2, and hence admit that the effects of ξ_1 and ξ_2 in this group are potentially confounded with ξ_3 to some unknown degree, or, if one is confident that the behavior of ξ_3 is the same in both groups, provide a fabricated statistical control for ξ_3 through construction of an indicatorless, or phantom, ξ_3 via a stacked model with equality constraints to group 1.[13] The indicated equality constraints make the phantom group 2 ξ_3 similar to the group 1 ξ_3 in three crucial respects. ξ_3 has the same variance, the same covariances, and the same effectiveness in both groups. *With these constraints in place, the effects of ξ_1 and ξ_2 on η_1 in group 2 are interpretable as effects that appear with ξ_3 controlled for the same degree of correlation and effectiveness it displayed in group 1.* That is, ξ_3 is now controlled in group 2, though it is acknowledged that this control reflects the assumed similarity of the action of ξ_3 in the two groups.

If, as in Figure 5.2B, two or more latent causes are missing in group 2, this can be modeled by a direct extension of the procedures used in Figure 5.2A.

If each group is to include one variable missing from the other one can apply a similar procedure, but there is an extra wrinkle. Figure 5.2C illustrates a model in which ξ_2 is measured in group 2 but not group 1, and ξ_3 is measured in group 1 but not group 2.[14] The between-group constraints are entered as before, but the wrinkle arises because the covariance between ξ_2 and ξ_3 does not make itself apparent in the indicators for either group. This forces the researcher to resort to a sensitivity analysis. That is, one does several stacked-model runs for each of a series of fixed ξ_2-ξ_3 covariances covering the range of reasonable values for this coefficient. Hence, one gains the ability to interpret all the effects of the ξ's on η_1 as effects *with all the other ξ's controlled (assuming similar ξ_2 and ξ_3 activity between the groups)* though these effects may or may not depend on the magnitude of the ξ_2-ξ_3 covariance, as indicated by the results of the sensitivity analysis.

Figure 5.2D depicts a slightly different context in which a phantom control variable is required and may be introduced by using a stacked model based on differing sets of indicators. In this instance, the researcher is interested in the effect of ξ_1 on η_2. The problem is that there are two very similar mechanisms that may be carrying effects from ξ_1 to η_2. The mechanisms are so similar that if one of them is measured, the other is disrupted and cannot be measured. That is, the procedure for measuring one of the mechanisms, here depicted as η_1, distracts the subject or in some way renders the operation of the other mechanism ineffective (so there is no direct effect of ξ_1 on η_2 in the upper or group 1 segment of Figure 5.2D). Hence, in group 1 where η_1 is measured, the "other mechanism" has been disrupted, so we know there is no "other" functioning mechanism that could provide a residual direct effect from ξ_1 to η_2.

The "other mechanism" is not explicitly represented as a variable in Figure 5.2D, but the presence of this mechanism appears specifically as the direct effect of ξ_1 on η_2 in the group 2 model. The effectiveness or existence of this other mechanism can be decided by looking to see if there is a null or substantial effect of ξ_1 on η_2 controlling for the mechanism η_1. But this is precisely the snag. How can one control for the η_1 mechanism if η_1 cannot be measured without disrupting the action of the alternate mechanism? The answer is to build a group 2 model in which the η_1 mechanism is operative, and hence whose effects are controlled for, but where η_1 has not been measured, so the possible "other mechanism" has not been disrupted. η_1 appears in the model, but only via the between-group constraints that artificially insert this model component into the group 2 model, rather than by direct measurement of η_1 in group 2. If the γ_{21} effect in group 2 is zero then there is no mechanism, other than η_1, via which ξ_1 influences η_2.

So the solution is yet again a stacked model based on unequal numbers of indicators, in this case the stacking of groups with and without the measured indicator(s) of η_1. By inserting the known effectiveness of the η_1 mechanism from the first group into the second group who have not had the alternative mechanism disrupted by the measurement η_1, *the researcher is able to control for the effectiveness of the η_1 mechanism in assessing the need for the other alternative mechanism (γ_{21}) in the second group.* The required intergroup constraints are straightforward. β_{21}, γ_{11}, and ψ_{11} must be equated to make η_1 function the same way in group 2 as it does in group 1. The test of the need for, or existence of, the alternative mechanism after controlling for the η_1 mechanism is to see if γ_{21} is significant in the second group.[15]

5.1.3 Testing the Sufficiency of Known Mechanisms

The discussion of Figure 5.2D above is oriented toward highlighting the need for the η_1 variable as a control variable in assessing the effectiveness of a possible alternative mechanism. This example could also have been described as an investigation of the sufficiency of a particular mechanism to account for a causal connection. Specifically, is η_1 sufficient to account for the entire effect of ξ_1 on η_2 or is some other mechanism necessary?

This slight shift of wording alerts us to an entirely different context in which stacked models based on differing sets of indicators can be applied.

Figure 5.3 Stacking to Assess the Sufficiency of Mechanisms.

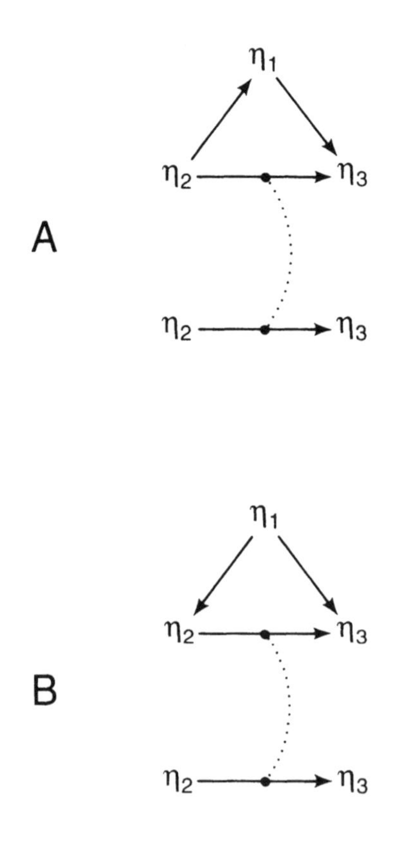

Imagine that in one group, say group 2 in Figure 5.3A, there is some causal mechanism that connects the concept η_2 to η_3. We know that there is an additional mechanism, containing η_1, that is operable in the other group, namely group 1. This mechanism could not possibly exist in group 2, so we have no reason to introduce an η_1 model segment into group 2. That is, unlike in Figure 5.2D, we know that the η_1 mechanism is applicable to only group 1 because we intervened to introduce the mechanism η_1 to counteract or supplement the usual effect of η_2 on η_3, or because the component η_1 exists in the context of group 1 but could never appear in the context of group 2. For example, η_1 may be an employment-related variable which is "not applicable" to the nonemployed in group 2.

By stacking these groups together and equating β_{32} between the groups, we can examine whether the η_1 mechanism is sufficient to account for the differential effectiveness of η_2 at producing η_3 in the two groups. That is, whether the occupational characteristic is sufficient to explain why the groups differ in the effectiveness of η_2 at producing η_3. If a stacked model without the β_{32} constraint works well, while the model with this constraint fails, we would conclude that there must be some mechanism other than η_1 which continues to differentiate the groups. In this case, the η_1 mechanism is insufficient to reconcile the group differences, so some additional, as yet unknown, mechanism must be operative in one or the other of the groups.

The issue is not whether the effects of η_2 on η_3 are the same in both groups. We know they are not because the η_1 mechanism is operative in only group 1 and this mechanism will never be present in group 2. The issue is whether the effects connecting η_2 to η_1 and η_1 to η_3 can be adjusted sufficiently in group 1 to permit the remaining effectiveness of η_2 on η_3 to be modeled as equivalent in both groups.

Note specifically that the effects of η_2 on η_1 and η_1 on η_3 would not be identical in stacked models with and without an equality constraint on the effect of η_2 on η_3. This means that attempting to assess the equivalence or nonequivalence of the effects of η_2 on η_3 is more than an issue of the statistical significance of the difference between the magnitudes of β_{32} in group 1 and group 2. There will be some flexibility in the estimates of the effects connecting η_1 to η_2 and η_3, and this flexibility permits some flexibility in the group 1 estimate of β_{32} which would be overlooked by a test that merely compared the estimated magnitude of β_{32} in the two groups. In moving from a stacked model with no β_{32} equality constraint between the groups to a stacked model with this constraint, the changes in the effects leading to and from η_1 will assist in absorbing some of the difference between the groups' β_{32} estimates in the stacked but unconstrained model.[16]

Whether that additional mechanism is intervening as in Figure 5.3A, or acts as a spurious cause as in Figure 5.3B, makes little difference. In both cases we are enquiring about whether some particular mechanism (η_1) is sufficient to account for the observed differences between the groups. And in both cases we will need a stacked model based on differing sets of indicator variables to investigate the sufficiency of the available explanation because η_1's indicators are available in only one of the groups.

5.1.4 Integrating Model Segments from Diverse Data Sets

Many of the pairs of models in Figures 5.1, 5.2 and 5.3 were intentionally represented as being similar with respect to both the data-gathering method and the form of the theoretical models for the two stacked groups. This simplified the presentation by focusing the discussion on the few relevant points of dissimilarity. It may often be reasonable to construct stacked models from groups that are subsets of a larger sample studied with a single method, but there is no formal demand that the stacked model segments arise in this way. Nor do the stacked model segments need to share the same general model form. All of the above issues can be investigated using different data sets and different model forms, as long as the stacked model segments share enough, similarly scaled, variables to permit constraints that address the relevant substantive issue.

The models in Figures 5.4 and 5.5 are designed to lessen the implicit dependence on similar model segments and data sets by illustrating some instances where the model segments are sufficiently different to make it easy to imagine these as arising from different studies. The models in these figures have the statistical control of otherwise unavailable variables, or the sufficiency of particular causal mechanisms, at their core. They do not introduce any new accomplishments beyond those discussed above. These models curb our urge to think of stacking as primarily a procedure for reducing sampling or modeling biases, and they alert us to the fact that stacked models can be applied in instances well beyond the examination of nearly parallel studies.

Figure 5.4A depicts two nonstacked models investigated by different researchers. These models contain two similarly measured variables, η_2 and η_4. Imagine that the data for the lower model in Figure 5.4A was reported in the literature some years ago, and that you and I are the researchers currently investigating the top model in some secondary data analysis. To our chagrin the model fails to match up with the data, and in particular it fails because the observed covariance between η_2 and η_4 is stronger than our model can be made consistent with. To make the model implication for this covariance stronger, larger estimates for β_{32} and β_{43} are required, but the observed covariances between η_2 and η_3, and η_3 and η_4 are too weak to permit the strengthening of these effects, so the model fails.

We are stymied until helpful-you remembers that someone else had once investigated the connection between η_2 and η_4, and found η_5 to be a common cause of these variables. We might now cite the reference for the bottom model as providing η_5 a possible explanation for the failure of our top model, but reviewers would be unlikely to believe us. From their perspective, appealing to η_5 would be mere handwaving to other variables as being necessary to salvage our ill-fitting top model.

Knowing how to stack models based on differing sets of indicator variables, we decide to tighten up our paper by demonstrating conclusively and specifically that the unavailability of η_5 in our data set is a complete and sufficient explanation for the ill fit of our top model. We will do this by integrating our model with the η_5 model, into a stacked model which hopefully will fit well in all its segments, including the segment corresponding to our originally ill-fitting model. We recognize that since η_1 and η_3 have no indicators in the second group, and η_5 has no indicators in the first group, we are probably into stacking models based on differing sets of indicators.

Figure 5.4 An Example of Stacking to Combine Model Segments.

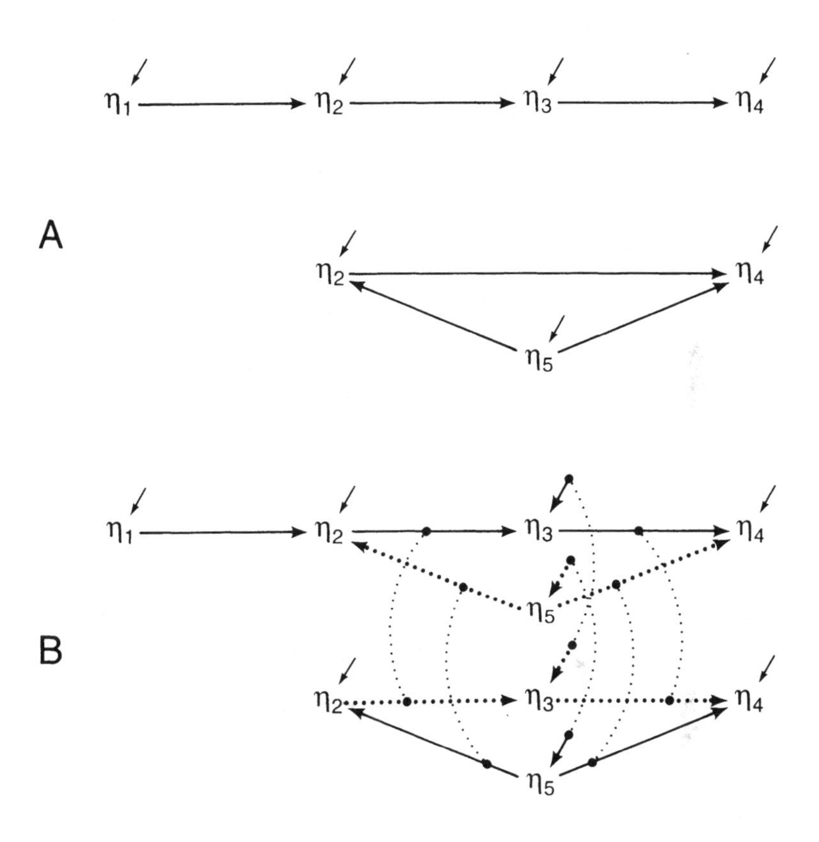

β_{32}, β_{43}, β_{25}, β_{45}, ψ_{33} and ψ_{55} are equal between the groups.

Figure 5.4B illustrates how we integrate the two model segments and data sets. We find we must include an indicatorless or phantom η_3 in the bottom model, and an indicatorless or phantom η_5 in the top model,[17] so that we can proceed to enter the indicated constraints. It comes as a pleasant surprise to find that this stacking is to the mutual benefit of both the top and bottom models. The top model benefits because we had correctly anticipated that the common cause η_5 provides exactly the proper amount of additional covariance between η_2 and η_4 to make this model segment fit with the data. And, assuming the data for the bottom model are also well fit, the bottom model, now model segment, benefits from the demonstration that η_3 provides precisely the proper amount of connection between η_2 and η_4 to replace the original β_{42} estimate in the unstacked lower model.

This leaves us in a much improved position. We need not write a paper guessing at the reason for the ill fit in our failing model. We can now report a well-fitting model arrived at through the integration of two separate data sets. Furthermore, our faithfulness to the form of both the original models substantively contributes to both the models. Would you as coauthor prefer to report a well-fitting model that synthesizes and enlightens two initially distinct theoretical model segments, or a model that fails, and a guess as to the reason for the failure?

The models in Figure 5.5 illustrate another stacking of dissimilar models, again intended to highlight that stacked groups need not come from parallel data sets, as long as the variables in the constrained portions of the model are, or can be made to be, usefully comparable.[18] We imagine the first study looked at the causal connections among a set of variables (the top group in Figure 5.5A). The second study examined only two of the variables present in the first study (η_1 and η_2) but it examined these in another context, namely with η_5 operative and measured (the lower group in Figure 5.5A). Figures 5.5A through C depict three ways of entering constraints between the stacked groups, and each of these ways corresponds to slightly differing theoretical assertions about the behavior of the variables in these models.

Figure 5.5A proposes that some intervention or circumstance has activated concept η_5 in the context of group 2, whereas η_5 is known or postulated to be inactive in the context of group 1. Comparing stacked models with and without the indicated equality constraint in Figure 5.5A can be used to assess whether the effect of η_1 on η_2 is the same in the two groups despite this context difference.[19]

Note that the compromise estimate of η_1's effect on η_2 in Figure 5.5A will not be the average of the two estimates when the models are run individually. One group may contain many more cases, and hence its estimate may more strongly resist change when arriving at the compromise estimate. And/or the equated coefficient in one of the group's models may be strongly overidentified, which implies that this group's estimate will more strongly resist

Figure 5.5 Another Example of Stacking to Combine Model Segments.

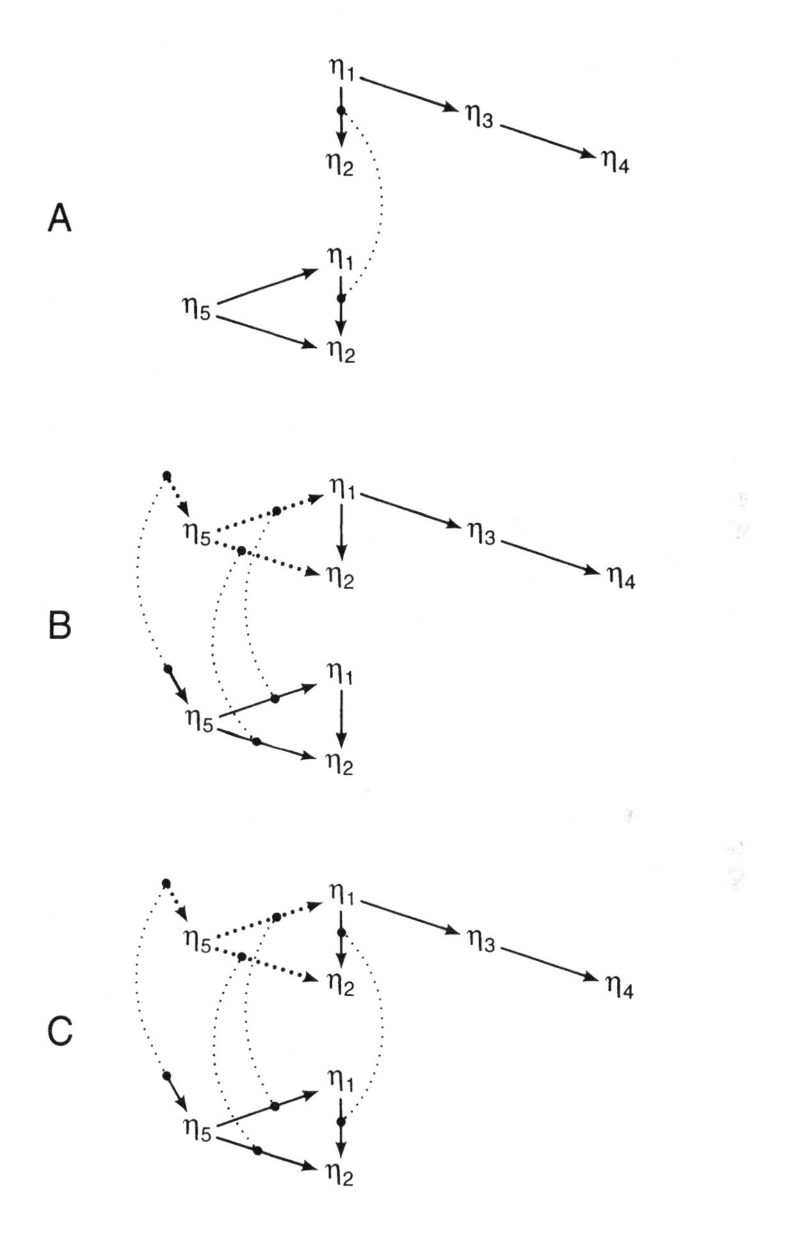

change because any change translates into increases in numerous residuals. That is, if the form of one of the models crucially involves this particular coefficient estimate in many more implied covariances, this group's estimate will resist change and hence will more strongly control the resulting compromise estimate. These competing sources of resistance in determining the compromise constrained estimate imply that one could not obtain a justifiable compromise estimate without estimating the constrained stacked model.

The stacked model in Figure 5.5B postulates that while η_5 was only measured and investigated in group 2, it is an active confounding factor in the context of group 1. Here the constraints create an indicatorless η_5 in group 1 that permits an adjustment[20] for η_5's impacts in the group 1 model, thereby permitting a more justified assessment of all the estimates in the group 1 model.

The model in Figure 5.5C is similar to the model in Figure 5.5B but it also constrains the effect of η_1 on η_2 to be equal between the groups. This corresponds to postulating both that η_5 was actively confounding the group 1 observations, and that η_5 is the sole reason the effect of η_1 on η_2 differed between the two groups. The Figure 5.5C model stops just short of claiming that the effect of η_1 on η_2 is identical in the two groups. This model does not postulate that ψ_{11} is equal in the two groups, so η_1 may have radically different variances in the two groups. This has implications for the amount of variance explained[21] in η_2. It also has implications for η_1's contribution to the spurious component of the covariance between η_2 and η_3.[22]

In sum, depending on the substantive context, one might wish to obtain a compromise estimate of the effect of η_1 on η_2 (Figure 5.5A), separate estimates of this effect with adjustment for the confounding behavior of another variable (Figure 5.5B), or a compromise estimate after adjustment for confounding (Figure 5.5C), all of which require stacked models with differing numbers of indicators.

In the belief that at least one of the modeling options discussed above has tempted you to try a stacked model with differing numbers of indicators, we now turn to the details of implementing the procedure in the context of a real example.

5.2 An Example: Stress, Drinking, and Employment Status

The model examined in this section is depicted in Figure 5.6 and is designed to illustrate how stacked models with unequal indicators can identify an otherwise underidentified model (Section 5.1.1). The data come from the 1993 All Alberta Survey,[23] and the codings for the relevant variables appear in Figure 5.7. The substantive aim of the example is to determine the causal

Figure 5.6 Stress and Moderate Drinking: An Example of Stacking to Attain Identification.

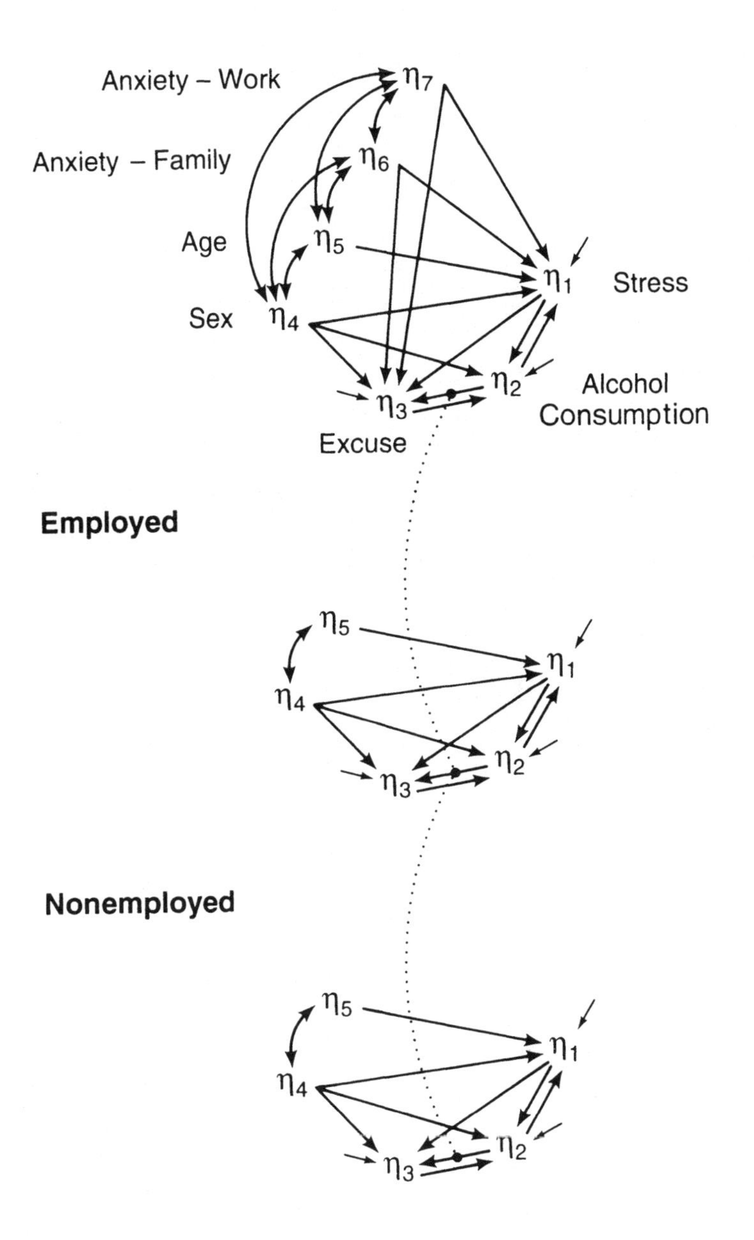

Figure 5.7 Indicators, Codings, and Assessed Proportions of Error Variance for the Stacked Drinking Model.

Concept	Assessed Error in Indicator	Questionnaire Item and Coding
Stress	5%	Would you describe your life as: very stressful (4) somewhat stressful (3) not very stressful (2) not at all stressful (1) no response (missing).
Alcohol	10%	On average how many drinks per week do you have (number of drinks per week coded to a maximum of 7) with "drink" defined as: One bottle of beer or glass of draft beer; one glass of wine; or one shot or mixed drink with hard liquor. To restrict the sample to moderate drinkers, non-drinkers (those who had not drunken alcohol in the previous 12 months) were coded as missing, and heavy drinkers (those reporting 8 or more drinks per week) were also coded as missing. Respondents drinking "once or twice a month" or "less than once a month" were not asked their frequency of drinking but were coded 1.0 and 0.5 respectively as a reasonable approximation of their frequency of drinking.
Excuse	5%	Do you drink more when under stress? Yes (3) don't know (2) no (1). Nondrinkers and no response (missing).
Sex	1%	Male (1), Female (2).
Age	1%	In years, with minimum 18. No response (missing).
Anxiety-Work	5%	My work responsibilities cause me anxiety at home. Strongly agree (7) to strongly disagree (1). Asked of employed only. Not working, don't know, and no response (missing).
Anxiety-Family	5%	My family responsibilities cause me anxiety at work. Strongly agree (7) to strongly disagree (1). Asked of employed only. Not working, don't know, and no response (missing).

connections between stress and alcohol use among moderate[24] drinkers of differing employment status. In particular, *does stress have the same causal impact on alcohol consumption among moderate drinkers who are employed, nonemployed* (those with no paying job but who are not seeking a job, such as homemakers and the retired), *and unemployed* (those with no paying job who are looking for work)?

Stress might influence alcohol use, and alcohol use might influence stress, so we are confronting a reciprocal relationship in the employed, nonemployed, and unemployed groups. To identify the coefficients comprising the reciprocal relationship in any one group, we need to locate a unique cause of one of the reciprocally connected variables, in this case a unique cause of alcohol consumption or a unique cause of stress. To break the symmetry of the reciprocal relation we need to locate a variable that is more intimately connected to one rather than the other of the reciprocally related variables.

But matters are even more complex. Moderate drinking is done in a social context, and that context introduces a second reciprocal relationship. Alcohol consumption occurs in a context of interpersonal relations which must be managed, and that managing may involve the use of rationalizations or excuses for drinking. Those with an available excuse/rationalization might consume more alcohol, and those consuming more alcohol might have more need of an excuse, so these reciprocal relations are also part of the model.

The excuse of interest is using "stress as a reason for drinking." We can not accept the respondents' assertions of whether or not they drink more when stressed as necessarily veridical descriptions of the complex causal world in which the respondents are located. We simply have no way to support the assertion that respondents can spontaneously and naively disentangle complex reciprocal causal networks. Hence, we must treat this not as a true description of a causal impact, but merely as another response made in a social context and whose causal origins are to be determined by our investigations. Subsequently, the model postulates that both real stress and alcohol consumption might influence the use of stress as an excuse, and the excuse might be reciprocally connected to actual alcohol consumption. Together, these effects introduce a second reciprocal relationship, and a loop that circulates from stress to excuse to alcohol consumption and back to stress, so we face a substantial task in attempting to identify these relationships.

We turn to the data set and find we are in trouble for two of the three employment groups. For the unemployed and nonemployed the only other relevant variables in the data set are the respondent's sex and age, whose effects are anticipated to be as diagrammed in Figure 5.6. Sex is of no help in our search for a unique cause since it is likely to influence all three of the reciprocally related variables. Age is expected to act as a unique cause of stress in all three groups, so the stress-alcohol effects are probably identified, but the excuse-alcohol reciprocal effects have no unique cause to break their symmetry, and the stress-excuse-alcohol loop can only make matters worse. Even a fine-tooth comb is unable to locate any other useful unique causes of excuse or alcohol in the nonemployed and unemployed groups.

The good news is that the employed individuals were asked two extra questions that might act as causes of excuse and stress, but not of alcohol consumption directly. Agreeing that "my work responsibilities cause me anxiety at home" (bringing work problems home) or that "my family responsibilities cause me anxiety at work" (being hampered by family responsibilities) are reports of conditions that are likely to produce stress in the respondent. Via stress, these might influence alcohol consumption, but these should not lead *directly* to alcohol consumption without going through stress. Hence, the work/family problems variables can act as unique causes which identify the stress-alcohol reciprocal effects in the employed group.

Similarly, the existence of anxiety-producing work and family responsibilities may also lead to an increased propensity to use "stress as an excuse," thereby identifying the excuse-alcohol reciprocal effect in the employed group.

So the reciprocal effect between stress and alcohol is probably identified by the unique effect of age in all three groups, and the anxiety variables in the employed group, but the reciprocal relation between excuse and alcohol is probably[25] identified in only the employed group, namely for only those individuals for whom the work/family questions could be asked.

Assuming the model for the employed persons is identified, is there any way to use this identified model to assist in identifying the models for the non- and unemployed? In particular, are either of the excuse-alcohol effects likely to have the same magnitude (to be equally effective) in all three groups? Constraining such an effect to be equal in all three groups should provide an additional model constraint focused on the problematic reciprocal effects between excuse and alcohol in the non- and unemployed groups. That is, we should be able to identify these reciprocal effects if we are willing to defend the claim that one of the effect estimates from the employed group should equal the corresponding effect in the non- and unemployed groups.

Fortunately, there is such an effect. We would anticipate that the effect of alcohol consumption on the propensity to use stress as an excuse should not change between the context of being employed, nonemployed and unemployed. The groups might differ in ways that make stress more or less connected to using stress as an excuse (β_{31}), or in ways that make stress more or less connected to alcohol use, and subsequently to differential use of stress as an excuse ($\beta_{21}\beta_{32}$), but both of these are different than arguing that alcohol use is more or less effective at producing the use of stress as an excuse in the different groups (β_{32}).

Then comes the crunch. How does one set up the stacked model to incorporate the equality constraint on β_{32}, the effect of alcohol on excuse, if LISREL will not accept different numbers of indicators in the stacked groups? The work/family anxiety variables in the employed group are necessary to identify the model but the nonemployed and unemployed groups do not have comparable indicators, and hence are based on smaller covariance matrices. The trick is to fool LISREL by creating some *filler variables* that do nothing in the model yet which trick LISREL into thinking that we have two additional indicators in the unemployed and nonemployed groups, and hence an equal number of indicator variables in all the groups.[26]

From the input covariance matrices for all three groups (Figure 5.8) you will see that the last two rows of the data matrices for the unemployed and nonemployed groups are rigged. The two *filler indicators* added to these groups are presumed to be independent of, and hence uncorrelated with, all the other variables in the model. And these filler indicators are specified as having unit variances, so the matrix does not become singular, as it would if an entire row of zeros was entered.

LISREL will now accept the data, but it will still not produce any estimates. Just as a row of zeros in the data produces a singular matrix which keeps LISREL from functioning, a row of zeros in the Σ matrix also results in a singular matrix, and hence also halts LISREL because this also makes the likelihood function unevaluatable. Thus, unless our model provides some model-implied variances for these filler indicators, LISREL will still refuse to provide maximum likelihood estimates.

The LISREL command lines provided in Figure 5.9 illustrate how to avoid a zero row in Σ by specifying *filler concepts* with unit variances corresponding to each of the *filler indicators*. Each filler concept is forced to have unit variance, but to be otherwise independent of all the other components of the model. This results in Σ having filler rows that are entirely zeros except for the unit variances, in a pattern that exactly matches the pattern of zeros and ones in the input data. A fixed 1.0 ψ provides a variance of 1.0 for each filler concept. The λ linking the filler concept to the filler indicator is fixed at 1.0 and the corresponding θ is fixed at 0.0 so that all, and only, the variance of the filler concept is conveyed to the filler indicator. The filler concepts are kept independent of all the other concepts in the model by permitting no other effects into or out of the filler concepts. This makes the indicators of the filler concepts independent of the other indicators in the model, as required by the zero covariances in the data matrices.[27]

This specification forces the filler variable rows of the data and Σ matrices to correspond exactly. Consequently, the filler rows of the data and Σ matrices do not influence LISREL's iterative search for the maximum likelihood estimates for the coefficients in the model. There is no way for any of the

Figure 5.8 Listwise Covariance Matrices for the Groups.

Employed (N=611)

	stress	alcohol	excuse	sex	age	anxietyF	anxietyW
stress	0.486						
alcohol	0.042	3.175					
excuse	0.051	0.272	0.434				
sex	0.016	-0.245	-0.019	0.250			
age	-0.638	0.917	0.046	0.019	115.914		
anxietyF	0.349	-0.323	0.012	0.089	-1.274	3.127	
anxietyW	0.484	-0.211	0.143	0.025	-2.897	1.351	3.844

Nonemployed (N=223)

	stress	alcohol	excuse	sex	age	FILLER1	FILLER2
stress	0.685						
alcohol	-0.195	2.421					
excuse	-0.022	0.269	0.316				
sex	0.064	-0.162	-0.026	0.239			
age	-4.847	4.735	0.683	-1.884	383.897		
FILLER1	0.000	0.000	0.000	0.000	0.000	1.000	
FILLER2	0.000	0.000	0.000	0.000	0.000	0.000	1.000

Unemployed (N=67)

	stress	alcohol	excuse	sex	age	FILLER1	FILLER2
stress	0.644						
alcohol	-0.163	2.389					
excuse	0.028	0.149	0.417				
sex	0.080	-0.122	-0.015	0.249			
age	-1.720	0.741	-1.172	0.806	177.089		
FILLER1	0.000	0.000	0.000	0.000	0.000	1.000	
FILLER2	0.000	0.000	0.000	0.000	0.000	0.000	1.000

model's free coefficients to improve upon the already perfect match between the filler rows of the Σ matrix and the filler rows of the data matrix **S**. The only less than perfect fit between the data **S** and Σ matrices occurs in the portions of the matrices corresponding to the real variables, and hence it is these real portions of the data and Σ which drive the estimation procedure.

To check on the basic assertion that it was indeed the equality constraint to the employed group that identifies this model, I began by running a model that was identical to the Figure 5.6 model except that β_{32} was constrained to be equal in only the non- and unemployed groups while it was permitted a separate estimate in the employed group. It was anticipated that the β_{32} coefficient would be unidentified in both the non- and unemployed portions of this model, and hence that this stacked model would not be identified. This was in fact the case. LISREL quit after 207 iterations, complaining of severe problems, possibly underidentified coefficients, and producing wild values, standardized β's greater than 1.0, and no modification indices, standard errors or *T* statistics.

In contrast, constraining β_{32} to be equal in all three groups, instead of just two groups, resulted in convergence in 42 iterations, full output, and no unanticipated[28] error messages. So it appears that we were indeed able to identify the coefficients in the non- and unemployed groups through use of a single equality constraint to the group having the additional indicators. The estimates for the effects among the concepts in this model appear in Table 5.1.[29]

The model fits acceptably, after adjusting the degrees of freedom for the filled data entries. Given that the inclusion of filler variables artificially expands the smaller data matrices, and given that the degrees of freedom are the difference between the number of entries in the data matrices and the number of estimated coefficients (Joreskog and Sorbom, 1988:228; *Essentials*, 162, 284), it is understandable that the reported degrees of freedom are too large. Each variance and covariance in the filled row(s) of the data matrices unjustifiably adds one degree of freedom to LISREL's reported χ^2 test. The perfect fit to the filled rows is not accomplished by the action of a substantively correct model matching up with real data, but by inserting fictitious data and designing some fixed model components to perfectly match with this fiction. The artificially inflated degrees of freedom result in the reported χ^2 probability being too large because the correct χ^2 value is assessed relative to the wrong χ^2 distribution, namely the distribution for the wrong degrees of freedom.

These problems can easily be corrected by either subtracting the total number of filled variances and covariances from the usual χ^2 degrees of freedom, and looking up the χ^2 probability in a χ^2 table. Or one can use a *DF=-#* command on LISREL's output line to reduce the degrees of freedom in versions 7 or later of LISREL.[30] If the correct degrees of freedom are obtained using the *DF=* command, LISREL correctly calculates the χ^2 probability.

Figure 5.9 LISREL Command File for the Stacked Model with Filler Variables.

```
 1  0  title 'STACKedDRINKING 3 grps FILLER VARS new model LSTACK.DR7'
 2  0  file handle #8/name='DRINKD2.EMP'
 3  0  file handle #9/name='DRINKD2.NON'
 4  0  file handle #10/name='DRINKD2.UN'
 5  0  INPUT PROGRAM
 6  0  NUMERIC A
 7  0  END FILE
 8  0  END INPUT PROGRAM
 9  0  USERPROC NAME=LISREL
10  0  STACKED DRINKING - EMPLOYED                                  LSTACK.DR7
11  0  DA NI=7 NO=611 MA=CM  NG=3
12  0  CM UN=8 FU FO
13  0  (8F10.7)
14  0  LA
15  0  'stress' 'alcohol' 'anxietyW' 'anxietyF' 'excuse' 'sex' 'age' /
16  0  SE
17  0  'stress' 'alcohol' 'excuse' 'sex' 'age'
18  0  'anxietyF' 'anxietyW' /
19  0  MO NY=7 NX=0 NE=7 NK=0 LY=FU,FI BE=FU,FI PS=SY,FI TE=SY,FI
20  0  VA 1.0 LY(1,1) LY(2,2) LY(3,3) LY(4,4) LY(5,5) LY(6,6) LY(7,7)
21  0  FR         BE(1,2)          BE(1,4) BE(1,5) BE(1,6) BE(1,7)
22  0  FR BE(2,1)         BE(2,3) BE(2,4)
23  0  FR BE(3,1) BE(3,2)          BE(3,4)         BE(3,6) BE(3,7)
24  0  FR PS(1,1)
25  0  FR       PS(2,2)
26  0  FR               PS(3,3)
27  0  FR                         PS(4,4)
28  0  FR                         PS(5,4) PS(5,5)
29  0  FR                         PS(6,4) PS(6,5) PS(6,6)
30  0  FR                         PS(7,4) PS(7,5) PS(7,6) PS(7,7)
31  0  VA   .02429   TE(1,1)
32  0  VA   .31748   TE(2,2)
33  0  VA   .021715 TE(3,3)
34  0  VA   .002498 TE(4,4)
35  0  VA  1.159144 TE(5,5)
36  0  VA   .156325 TE(6,6)
37  0  VA   .192185 TE(7,7)
38  0  ST   .5 PS(1,1)
39  0  ST 3.0 PS(2,2)
40  0  ST   .4 PS(3,3)
41  0  ST .25 PS(4,4)
42  0  ST 110. PS(5,5)
43  0  ST 3.7 PS(6,6)
44  0  ST 3.0 PS(7,7)
45  0  OU
46  0  DRINKING STACKED MODEL --- NON-EMPLOYED group
47  0  DA NI=7 NO=223 MA=CM
48  0  CM UN=9 FU FO
49  0  (8F10.7)
50  0  LA
51  0  'stress' 'alcohol' 'excuse' 'sex' 'age' 'FILLER1' 'FILLER2' /
52  0  SE
53  0  'stress' 'alcohol' 'excuse' 'sex' 'age' 'FILLER1' 'FILLER2' /
54  0  MO NY=7 NX=0 NE=7 NK=0 LY=FU,FI LX=FU,FI BE=FU,FI PS=SY,FI TE=SY,FI
```

Figure 5.9 Continued.

```
 55  0  VA 1.0 LY(1,1) LY(2,2) LY(3,3) LY(4,4) LY(5,5) LY(6,6) LY(7,7)
 56  0  FR          BE(1,2)         BE(1,4) BE(1,5)
 57  0  FR BE(2,1)          BE(2,3) BE(2,4)
 58  0  FR BE(3,1) BE(3,2)          BE(3,4)
 59  0  FR PS(1,1)
 60  0  FR          PS(2,2)
 61  0  FR                   PS(3,3)
 62  0  FR                           PS(4,4)
 63  0  FR                           PS(5,4) PS(5,5)
 64  0  VA  .034255 TE(1,1)
 65  0  VA  .24215  TE(2,2)
 66  0  VA  .01582  TE(3,3)
 67  0  VA  .002388 TE(4,4)
 68  0  VA 3.838971 TE(5,5)
 69  0  VA  .0000   TE(6,6)
 70  0  VA  .0000   TE(7,7)
 71  0  ST .65 PS(1,1)
 72  0  ST 2.2 PS(2,2)
 73  0  ST .3  PS(3,3)
 74  0  ST .25 PS(4,4)
 75  0  ST 380. PS(5,5)
 76  0  VA 1.0 PS(6,6)
 77  0  VA 1.0 PS(7,7)
 78  0  OU
 79  0  DRINKING STACKED MODEL --- UN-EMPLOYED group
 80  0  DA NI=7 NO=67 MA=CM
 81  0  CM UN=10 FU FO
 82  0  (8F10.7)
 83  0  LA
 84  0  'stress' 'alcohol' 'excuse' 'sex' 'age' 'FILLER1' 'FILLER2' /
 85  0  SE
 86  0  'stress' 'alcohol' 'excuse' 'sex' 'age' 'FILLER1' 'FILLER2' /
 87  0  MO NY=7 NX=0 NE=7 NK=0 LY=FU,FI LX=FU,FI BE=FU,FI PS=SY,FI TE=SY,FI
 88  0  VA 1.0 LY(1,1) LY(2,2) LY(3,3) LY(4,4) LY(5,5) LY(6,6) LY(7,7)
 89  0  FR          BE(1,2)         BE(1,4) BE(1,5)
 90  0  FR BE(2,1)          BE(2,3) BE(2,4)
 91  0  FR BE(3,1) BE(3,2)          BE(3,4)
 92  0  EQ BE(1,3,2) BE(2,3,2) BE(3,3,2)
 93  0  FR PS(1,1)
 94  0  FR          PS(2,2)
 95  0  FR                   PS(3,3)
 96  0  FR                           PS(4,4)
 97  0  FR                           PS(5,4) PS(5,5)
 98  0  VA  .032195 TE(1,1)
 99  0  VA  .23885  TE(2,2)
100  0  VA  .02085  TE(3,3)
101  0  VA  .002486 TE(4,4)
102  0  VA 1.770890 TE(5,5)
103  0  VA  .0000   TE(6,6)
104  0  VA  .0000   TE(7,7)
105  0  ST .65  PS(1,1)
106  0  ST 2.2  PS(2,2)
107  0  ST .3   PS(3,3)
108  0  ST .25  PS(4,4)
109  0  ST 175. PS(5,5)
110  0  VA 1.0  PS(6,6)
111  0  VA 1.0  PS(7,7)
112  0  OU ML NS AL TM=40
113  0  end user
```

Table 5.1 Maximum Likelihood Estimates and Standard Errors for the Stacked Model.[a]

	Employed	Non-employed	Un-employed
β_{12}	.157*	.181	.099
	.078	.152	.366
β_{14}	.183	.287	.413
	.093	.153	.277
β_{15}	-.003	-.014**	-.012
	.003	.003	.008
β_{16}	.078**	----	----
	.019		
β_{17}	.109**	----	----
	.016		
β_{21}	-.821	-.653	-.583
	.447	.393	1.190
β_{23}	1.415	1.549	.931
	1.589	1.369	1.246
β_{24}	-.828**	-.336	-.251
	.210	.274	.565
β_{31}	-.103	-.177	-.026
	.302	.227	.284
β_{32}	-.107†	-.107†	-.107†
	.244	.244	.244
β_{34}	-.177	-.138	-.107
	.234	.148	.205
β_{36}	-.014	----	----
	.021		
β_{37}	.053	----	----
	.032		

† Constrained to be equal in all three groups.
* Estimate exceeds 2.0 times its standard error.
** Estimate exceeds 3.0 times its standard error.
a The standard error appears below each estimate.

LISREL 6 reported the χ^2 for the stacked drinking model as 1.82 with 30 degrees of freedom and p=1.00. The filler variables introduced a total of 26 0.0 and 1.0 artificial data entries, so the true degrees of freedom should be 30 - 26 = 4 with the same χ^2 of 1.82 for a probability of about .77, which is acceptable.[31]

With most of the modeling issues behind us, we can examine some of the substantive stories told by the estimates. First, the effect of alcohol consumption on stress (β_{12}) is small and positive in all three groups. If anything, consuming additional alcohol produces slightly more stress among these moderate drinkers. There may be something other than the biologically calming action of moderate amounts of alcohol at work here. Some aspect of the social matrix surrounding the moderate alcohol consumption may be making minor[32] increases in consumption more stressing. If true, this makes "drinking to relieve stress" not only an excuse, but descriptively wrong, because this contradicts the actual null or stressing consequences of increased alcohol consumption among moderate drinkers.

A second observation is that increased stress tends to decrease the consumption of alcohol in all three groups (β_{21}). Each additional increment in stress (on a four-point scale) reduces alcohol consumption by just over half a drink per week in all three groups. It is possible that the additional responsibilities and financial or time pressures accompanying higher stress levels inhibit alcohol consumption. One may simply not have the time, money, or freedom to consume in the way one does when less pressed.

These estimates oppose lay thinking which sees increasing stress as leading to increased alcohol consumption, and increased alcohol consumption as decreasing stress. Among moderate drinkers at least, the reverse seems more likely. Increased stress leads toward abstinence, while increased drinking leads to more stress.

Then comes an observation that leads to considerable reevaluation. Each increment in the availability of "stress as an excuse" leads to consumption of about an additional drink per week in all three groups (β_{23}). While the sign of these effects seems reasonable, it is their insignificance that prompts the reevaluation. The insignificance to borderline significance of the β_{21} and β_{12} effects above are understandable given that these effects when standardized were in the range of .1 to .3, but the β_{23} effects seem more substantial. Indeed, when examined as standardized effects[33] these are in the .4 to .5 range. So why are they not significant? The answer is clearly that the standard errors of these coefficients must be large. But why should the standard errors be large if these are well identified? Did not the work/family variables make the corresponding reciprocal effect in the employed model well identified, and the equality constraint transfer this well-known coefficient to all the groups so that the other half of the reciprocal effect could be well determined?

The answer is neither yes nor no, but somewhat! The identification of the reciprocal effect of excuse on alcohol in the employed group was to be gained by the introduction of the unique causes of work and family problems. But if we examine the effects of these variables on excuse (namely β_{36} and β_{37}) we see that neither of these is truly significant. β_{36} is essentially zero, and if anything of the wrong sign, so family problems certainly did not function to assist in identifying the reciprocal effects. The T value for β_{37} is 1.654 so we might claim this as a borderline significant effect if we wished to assert a prior commitment to the direction of the effect, but this clearly remains marginal in either case. That is, neither of the two unique causes we were counting on to identify the reciprocal effects in the employed model is a particularly strong cause of excuse. This means that these variables exercise some, but not particularly strong control over the coefficients in the reciprocal effects, and we notice this as the large standard errors which make the moderately large β_{23} effects insignificant.

So this example turns out to be instructive in an unanticipated way. It reminds us of the fact that identification is not an all or none affair. It is a matter of degree. And clearly the estimates we have here are right on the borderline of failing, not because the logic of the process is wrong, but because the strengths of the necessary effects are borderline.

How would one proceed if this were a real model and not merely an example? There are several options. Given that identification problems are in general improved by requesting fewer estimates, especially in the difficult segments of models, one could begin by assessing whether even the tentative estimates provided by this model are sufficient to warrant some model simplification. For example, if we could eliminate β_{32} from all the groups this would eliminate the problematic reciprocal effect, and hence improve the identifiability of the model. We should not use the insignificance of this effect (T=-.440) as a guide because this insignificance may be forced by the inflated standard errors on the reciprocal effects. A proposal to delete this would depend more on the metric magnitude and sign of this effect, which might be viewed as reversed.

Another style of option is created by considering the effects leading out from sex (η_4). The effects leading from sex to excuse (β_{34}) are potentially of inexplicable sign, small, and insignificant. The insignificance here is unlikely to be an artifact of artificially inflated standard errors because this is not part of a looped or reciprocal effect. Hence this too is a candidate for elimination. The reason this is attractive as a possible elimination is that with β_{34} gone, sex would be left with a unique effect to alcohol so that it would assist in identifying the alcohol-excuse reciprocal effects.

A third option is to drop the effect from stress to excuse (β_{31}). If this effect were eliminated, stress (η_1) would act as unique cause of alcohol in the alcohol-excuse relationship and hence assist in identifying the more problematic of the reciprocal effects. Naturally, the choice from among these options, or even moving to a sensitivity analysis for any of the model coefficients, should be based on the researcher's hunches regarding which of these would be the most justified.

Overall these borderline estimates and their standard errors might be sufficient to justify model revision, where those revisions might render all the groups solidly identified, with or without the constraints between the groups that necessitated our stacking of groups based on differing sets of indicators. Even though this particular stacked model is borderline in some respects, it nonetheless provides diagnostic information focusing the researcher's attention on specific new options and possibly even sufficient diagnostic information to justify some substantial revision to the original model. All in all, we have managed to gain estimates for effects that would otherwise have been unidentified in two of the three groups. The weakness of the effects from the unique causes in the employed group results in the standard errors for some of the effects being larger than we might like, but the model nonetheless displays considerable diagnostic utility.

5.3 Summary

Stacked, or multiple-group, LISREL models can be used to test for interactions, which increase control possibilities, and to estimate means and intercepts for the conceptual variables, which permit enhanced interpretations. Stacked models have been relatively infrequently utilized, but this is likely to change as LISREL-literate researchers realize how easy stacked models are to specify and as more reasons for using stacked models are discovered.

Traditional discussions of stacked models have been based on the assumption that the researcher has the same number of indicator variables in each of the stacked groups' data matrices. This chapter demonstrates that the requirement for equal numbers of indicators is unnecessary, and that a simple ploy that gets LISREL to accept stacked model segments based on differing numbers of indicator variables greatly extends the range of potential uses for stacked models.

In particular, the between-group constraints in stacked models can be used to identify otherwise underidentified models, whether that under-identification is in the structural or measurement portion of the model. The drinking example illustrates how the underidentified reciprocal effects in two of three groups can be identified by constraints to a third group which contains additional indicator variables. Indeed, with large data sets, one can imagine stacking several groups, where each of the additional stacked groups is intended to identify some particular portion of a seemingly hopelessly underidentified model.

Stacked models based on differing sets of indicator variables can also be used to control for unmeasured variables (Section 5.1.2), to test whether specific mechanisms are sufficient to explain the differences between groups (Section 5.1.3), and to integrate model segments from nonparallel data sets (Section 5.1.4). And still other uses are possible. For example, similar models might be useful in the context of reducing estimation bias through the use of remeasured, or partially remeasured, subsamples which result in replicate measures being available for only some of the cases (Allison and Hauser, 1991).

Notes

1. See Bollen (1989a:355-369), Joreskog and Sorbom (1989:227-244), or *Essentials* (277-286). The two separate models and the single large stacked model produce the same coefficient estimates unless one enters some constraints between the stacked groups. For example, an effect in one group might be constrained to be equal to, to be twice, or to be the negative of the effect in the other group. Such models may require the addition of phantom concepts as discussed in *Essentials* (199-202).

We use the term stacked rather than multisample because multisample seems to imply that different samples are required. In fact, the procedures are applicable with any nonoverlapping subgroups one wishes to designate within one's data set. This may be groups of males-females, old-young, rich-poor, religions A, B, and C, those agreeing versus those disagreeing with an item, those interviewed by telephone rather than in person, or interviewed by a male interviewer rather than a female interviewer. In fact, all of these styles of stacking could be used within a single sample.

2. This may be as simple as placing the two groups' model specifications into a single command file and specifying NG = 2 (for Number of Groups = two).

3. To test for equivalent structural effects, the B, Γ, and maybe Φ and Ψ matrices are constrained to be equal between the groups while the Λ or Θ estimates are permitted to differ between the groups. To test for equivalent measurement structure, one constrains the Λ and/or Θ estimate to be equal between the groups, while permitting the structural coefficients linking the concepts to differ between the groups. The test is a difference χ^2 test created from the stacked models with and without the indicated constraints. *Essentials* Chapter 9 discusses the specification and testing of such models.

4. These included complaints that "a selected variable was not found" (namely the seemingly "missing" variable in the group with the fewer variables), or that the "input data matrix is not positive definite" because a mysteriously arising row of zeros (and inexplicably labeled "CONST.") had been inserted into the smaller covariance matrix.

I invented a procedure for running stacked models with unequal numbers of input indicators, which circumvents these error messages, only to discover that this was a reinvention. An example using a similar strategy is provided by Allison (1987) and is cited later in the LISREL 7 manual (Joreskog and Sorbom, 1988:260). This example, however, is likely to be overlooked because it appears in a section on models with means, which are rarely used, and it concerns data missing from a random subsample, which might mislead one into thinking that the procedure is only applicable if one has a random sub-sample or if missing data is one's primary concern. Other uses of the procedure appear in the 1992 edition of the EQS manual (Bentler, 1992a:199), Gillespie and McDonald (1991), and Allison and Hauser (1991).

5. See Bollen (1989a:88-103) or *Essentials* (139-149).

6. This was incorrectly reported as an underidentified model in *Essentials* Figure 5.6, page 146. See Rigdon (forthcoming) for a discussion of the identification of this particular model. You can check for yourself that this is identified by estimating this model with an artificial 3*x*3 data matrix, as long as you use the fixed λ and θ procedure discussed in Chapter 1.

7. In fact, if there were no other between-group coefficients at issue, one could have accomplished nearly the same thing by estimating the top group alone, and then using one of the resulting reciprocal effect estimates as a fixed value in a separate run for the second group. These runs would not be exactly the same as the stacked run however. The fixed coefficient would have zero standard error in the second of the two separate runs, whereas it has a standard error in the stacked procedure, so differing significances of various effects (the other reciprocal, indirect and total effects) might arise. In practice, one is likely to prefer the stacked model because the similarity of the groups' model segments will arouse interest in other between-group constraints.

8. Or the equivalence of "left-pointing effects" since the same difference χ^2 should arise if one moves from either of the "saturated" group 2 models in Figure 5.1A or 5.1B to "the same" overidentified, or more constrained, model 5.1C in constructing the difference χ^2. See Joreskog and Sorbom (1988:216-217) or *Essentials* (163-167) on the use of the difference χ^2.

9. In this particular model this amounts to testing whether the variables at the two ends of the reciprocal relations have the same covariances in the two groups. If "the rest" of the models in which a component like Figure 5.1C is imbedded are not identical, this will not amount to a test of equal covariance between the groups.

Note also that this could not be done in two separate runs for the two separate groups as alluded to in note 7. Here the overidentification of the coefficients in the second group means that the second group has some leverage and can change the estimates appearing for the first group. Hence, this is not the same as simply applying the estimates from a separate run for the first group to a run for the second group.

10. See *Essentials* (247-253).

11. See *Essentials* (112-113).

12. I used artificial data sets and the stacked models in Figure 5.1 to check the above assertions about the Figure 5.1 models. You can check these models for yourself by creating similar but slightly different covariance matrices for two groups and then using these artificial data matrices to estimate whichever stacked model is of concern to you. The key evidence you would be seeking in the output would be signs of identification problems (*Essentials*, 139-142).

13. If controlling for the phantom ξ_3 is to alter the estimated effects of either ξ_1 or ξ_2 on η_1 in group 2, ξ_3 must have an effect on η_1 and be correlated with ξ_1 or ξ_2 in group 1. If ξ_3 was ineffective in group 1, or if ξ_3 was uncorrelated with ξ_1 and ξ_2 in group 1, there would be no need for using ξ_3 as a control variable. The reason for this parallels the discussion in *Essentials* (48) of the conditions required for a regression slope to change with the inclusion of an additional control variable.

14. The numbers of indicators in the groups may or may not be equal, depending on how many indicators there were for ξ_2 in group 2 and ξ_3 in group 1.

15. One could also accomplish this by entering a phantom η_1 with β_{21}, γ_{11}, and ψ_{11} fixed to the group 1 estimates into a nonstacked model for the second group. In practice, however, one would likely also want to include equality constraints on other model

coefficients such as γ_{22} and Φ_{21}, and it is these additional constraints that would necessitate a stacked unequal-indicator model.

16. If η_1 in Figure 5.3B had no indicators we could provide it fixed 1.0 effects to both η_2 and η_3 and use the variance of η_1 to control η_1's contribution to the covariance between η_2 and η_3 (e.g., Hayduk and Avakame, 1990/91). The stacked model could then be used to answer the question: How much of the covariance between η_2 and η_3 would have to be accounted for by some external source before one could reasonably maintain that η_2 has equal effects on η_3 in both groups? Under this scenario, η_1 has no indicators, so the stacked groups have an equal number of indicator variables but an unequal number of conceptual variables.

17. Which is not problematic given the discussion in *Essentials* (199-204).

18. We are not yet in a position to detail what it takes for model segments to be comparable. In most instances this would seem to require at least two variables being shared by the models (though even a single shared variable might be sufficient to force a desired variance equivalence), and a similar scaling of those variables between the groups (so the equality constraints are interpretable). Fixed λ values other than 1.0 in one of the groups might be helpful here.

Beyond this, one seems to be constrained primarily by two general concerns. First, the stacked model segments should be compatible with one another so that they do not claim mutually contradictory things about the sources of the shared indicators' covariances. Precisely what would constitute incompatible claims is not obvious. For example, it could be possible for η_1 to cause η_2 in one group, and η_2 to cause η_1 in the other, if the true relationship was reciprocal and one of the reciprocal effects was rendered inoperative in each of the groups. In this case no equality constraints would be entered to create reciprocal effects in either of the model segments, even though contrasting but still reasonable effect orderings appear in the model segments.

Second, the reasonableness of any particular stacking will depend on the substantive meanings of the variables and the consistency of those meanings between the model. If this is true, it will be difficult or impossible to develop cookbook rules for differentiating reasonable from unreasonable stacked models on the basis of model characteristics alone. Radically different fixed θ values, for example, could provide substantively different conceptualizations even though the models share the same model form by having fixed non-zero θ's.

19. The covariance between η_1 and η_2 would differ between the groups because η_5 contributes to this covariance in only group 2.

20. This is a mere adjustment because the created η_5 is not as good as a real η_5 with indicators. With real indicators, the covariances between the indicators of η_5 and the indicators of η_3 and η_4 would provide further information to assist estimation and testing.

21. See the preface figure, row three, column two, or *Essentials* (20).

22. See the preface figure, row three, column three, or *Essentials* (31).

23. This was a telephone survey of 1274 respondents in Alberta, Canada, in 1993. The details of the sampling and case weighting are available in Kinzel (1993). I thank Dr. Susan McDaniel for permitting us early access to several of the variables that she had placed on this survey. Copies of the AAS93 data as an SPSS system file may be obtained by writing to the Director of the Population Research Laboratory, Department of Sociology, University of Alberta, Edmonton, Alberta, Canada T6G 2H4.

24. Moderate is defined as those who do drink, but who do not consume more than one drink per day, on average.

25. "Probably" is used here to acknowledge that the identification of these segments of the model depends on substantial effects arising from the unique causes. If unique effects are postulated, but the corresponding effects in the real world are null, then the unique causes are not really causes and are of no help in identifying the reciprocal effects. They make no contribution to breaking the symmetry of the reciprocal effects, and if anything make matters worse by requesting estimation of useless null effects.

26. A parallel procedure using filler variables is required by EQS (Bentler, 1992a:198-200) though the details of what must be attended to during EQS model specification differ somewhat.

27. An alternative way of setting this up is to specify all of the 1.0 variance of the filler indicators as coming from fixed 1.0 θ variances while simultaneously excluding all error covariances to the filler error variables. See Allison (1987) and Joreskog and Sorbom (1988) for examples. One could even split the 1.0 variance between a concept (e.g., .6) and measurement error variance (.4) as long as all the parts of the variance are modeled as being independent of the other variables in the model. Splitting the variance can be used to avoid entire rows of zeros in the Θ, Φ, or Ψ matrices, and hence to avoid messages such as "Warning: The theta epsilon matrix is not positive definite" which is prompted by the zero Θ_ϵ variances in the current setup.

28. The error message that the Θ_ϵ matrix is not positive definite in both the non- and unemployed groups is unproblematic because the filler indicators are provided zero error variances and hence entire rows of the Θ_ϵ matrix are zero.

29. The measurement sections of the model are not presented because each concept has a single indicator, with the corresponding λ fixed at 1.0 and the corresponding Θ_ϵ fixed at the assessed proportions of error variance as indicated in Figure 5.7. See the input program in Figure 5.9 for the specific values. The relatively close match between our concepts and the indicators, and the obviousness of the reasons for the larger of the measurement error assessments mean there is little to be gained from a detailed discussion of the measurement structure in this particular model. The estimated concepts' covariances are nearly the same as the corresponding indicator covariances, and the concepts' variances nearly equal the indicator variances minus the specified error variance, so there is no need to repeat these either in Table 5.1.

30. When I used filler variables in LISREL 7 there was an unexplained adjustment to the degrees of freedom that corrected for the filled zero entries but not the filled 1.0 diagonals. This makes me recommend always doing a hand calculation of the degrees of freedom to determine the correction required.

31. "Acceptable" according to the standard presented on page 69. The perfectly fit artificial data entries also result in zero residuals and hence appear in the *Q-Q* plot as a vertical line of points near the center of the plot. The artificial zero residuals also downwardly bias the root mean square residual.

32. Only minor increases in consumption are possible because only those who drink to a maximum of one drink per day are included in this model.

33. Using the within group standardized solution suggested in *Essentials* (183-184) and implemented in LISREL 7.14.

Chapter 6

Tidying Up

This housekeeping chapter is intended to air two concerns and to organize and dust off the literature that has been accumulating while we were not looking. We begin with the coefficient of determination which has been reported from LISREL 5 onward. I have found no use for this coefficient, and I hope that explaining my dissatisfactions will either permit me to stash this in the statistical basement, or will prod someone into providing a tidy and useful interpretation.

Next we peek under the edge of some statistical fixtures and notice the dirt lurking just out of reach of the factor-analytic attachment to our Monte Carlo studies. Numerous Monte Carlo studies have polished our understanding of the statistical behavior of various estimators, tests, and indices but whether this is sufficient to guarantee statistical cleanliness is questionable because the factor-analytic attachment simply is not long enough to reach all the way to the real dirt of current research practice. The factor model was once, but is no longer, typical of the models being estimated. Hence, demonstrating the utility or deficiencies of specific statistics in the context of the factor model is no longer a convincing demonstration of the cleanliness of our real statistical house. In Section 6.2, I propose a motif model whose advanced cleaning power should be able to put a shine on even the filthiest of our statistical modeling problems.

Section 6.3 concludes our housekeeping by organizing the scattered literature into piles of neatly fit indices, categoric estimation, and multilevel models. So grab a broom and let's begin with the statistical cobwebs!

6.1 A Quibble over the Coefficient of Determination

The LISREL 7 manual includes the coefficient of determination as one of five things that researchers "should pay careful attention to" in assessing the adequacy of a model (Joreskog and Sorbom, 1989:40, and see Joreskog and Sorbom, 1981:I-37).[1] This admonition stands in stark contrast to my personal inability to find any substantive use for this coefficient. Permit me to sketch my concerns.

The definition of the coefficient of determination (*CoD*) parallels the traditional definition of R^2, or the proportion of explained variance in a dependent variable. For example, in a model containing an endogenous concept η_3, the proportion of explained variance in η_3 is calculated as 1.0 minus the proportion of unexplained variance in η_3, namely the variance of the corresponding error variable ζ_3 or ψ_{33}. This can be written as

$$R^2 \text{ for } \eta_3 = 1.0 - \frac{\psi_{33}}{\text{Var}(\eta_3)} . \qquad 6.1$$

If all of η_3's variance is error, the numerator and denominator of the fraction are equal, and R^2 becomes zero.[2] If there is zero error variance, the variance in the dependent variable is fully explained and R^2 becomes 1.0.

The formula for the coefficient of determination is patterned directly on the formula for R^2 but it refers to the full set of endogenous concepts and their corresponding error variances:

$$\text{CoD} = 1.0 - \frac{|\Psi|}{|\text{Cov}(\eta)|} . \qquad 6.2$$

Recalling that the determinant of a matrix is a single number calculated from the elements of the corresponding matrix,[3] we see that the coefficient of determination is 1.0 minus a fraction created by dividing something about the errors (namely their determinant) by something about the variances and covariances of the corresponding endogenous variables (also their determinant). Hence the coefficient of determination parallels R^2, but precisely what the determinant is doing is not yet clear.

Imagine a model with two η's and their corresponding ζ error variables. We will assume the η's are standardized and that the error variables are independent of one another, but we need not know any further details of the model's specification (how many ξ's there are, which η's they cause, or whether the η's are causally connected). The coefficient of determination for this two-η model is

$$\text{CoD} = 1.0 - \frac{\begin{vmatrix} \psi_{11} & 0 \\ 0 & \psi_{22} \end{vmatrix}}{\begin{vmatrix} 1.0 & \text{Cov}(\eta_1\eta_2) \\ \text{Cov}(\eta_1\eta_2) & 1.0 \end{vmatrix}} . \qquad 6.3$$

From this, it is clear that the estimates of ψ_{11}, ψ_{22}, and $\text{Cov}(\eta_1\eta_2)$ control the numerical magnitude of the coefficient of determination. Recalling that the determinant of a 2x2 matrix is the product of the elements on the main diagonal minus the product of the elements on the other diagonal,[4] we can express the coefficient of determination for the model with two standardized η's and independent errors as

$$\text{CoD} = 1.0 - \frac{\psi_{11}\psi_{22} - 0.0\text{x}0.0}{1.0\text{x}1.0 - \text{Cov}(\eta_1\eta_2)\text{Cov}(\eta_1\eta_2)} . \qquad 6.4$$

The numerator of the fraction informs us that as the error variances increase,[5] the coefficient of determination decreases for any given value of $\text{Cov}(\eta_1\eta_2)$. This parallels the behavior of R^2. And, if either error variance becomes zero the coefficient of determination becomes 1.0, no matter how large the other error variance. So, according to the coefficient of determination, explaining one η completely and the other η not at all is as good as explaining both η's completely, because both lead to a *CoD* of 1.0. Interesting! Useful?

Now, what about the $\text{Cov}(\eta_1\eta_2)$, which would be a correlation since the η's are standardized? The above equation tells us that, for any particular values of ψ_{11} and ψ_{22}, the larger the covariance between η_1 and η_2 (positive or negative) the smaller the denominator, and hence the smaller the overall coefficient of determination. Given that we prefer a larger value for the coefficient of determination, this means that a model implying a smaller $\text{Cov}(\eta_1\eta_2)$ will therefore be preferred over a model implying a larger $\text{Cov}(\eta_1\eta_2)$, no matter whether that covariance arises from a direct effect, reciprocal effects, or dependence on common causes. That is, the coefficient of determination is telling us to prefer models in which the endogenous concepts are less strongly correlated, irrespective of the sources of that covariance. For any given size of error variances, the coefficient of determination is encouraging us to favor models that accomplish this explanation without implying any covariance between η_1 and η_2, as would happen if η_1 was caused by variables that were independent of the causes of η_2.

Why should we prefer models whose endogenous variables require separate explanations over models in which there is substantial covariance between the endogenous concepts because, say, one of the endogenous variables causes the other? This does not seem to make much sense. It is encouraging us to build models whose explanatory components are separate and distinct rather than interconnected. Why should we prefer to build segmented rather than unified theories of the behavior of two endogenous concepts?

Is it possible that this aspect of the behavior of the coefficient of determination is reasonable for only some models, and not all models? Specifically, is it reasonable for the factor model which provided LISREL's roots? Unfortunately, even in the context of the factor model, the behavior of the $\text{Cov}(\eta_1\eta_2)$ term does not seem to make sense. In factor analysis the basic source of covariance is dependence on common causes, with larger covariances arising from stronger loadings on any one common cause or stronger loadings on two correlated causes. From this perspective, the preference for a low $\mathrm{Cov}(\eta_1\eta_2)$ in the coefficient of determination is encouraging us to prefer strong but separate (unique and independent) factors. This is clearly at odds with the factor-analytic aim of seeking a few common causes of many items!

Let us leave this concern hanging and see if we can find a justification for a related concern, namely a rationale for why we should prefer highly covarying error variables. Imagine now that the error variables attached to our two endogenous concepts are not independent. This means that the zero off-diagonal entries in the ψ matrix above should be replaced with some specific covariance ψ_{12}. Consequently the two zeros in the numerator of the fraction in Eq. 6.4 would be replaced by $\psi_{12}\psi_{12}$:

$$\mathrm{CoD} = 1.0 - \frac{\psi_{11}\psi_{22} - \psi_{12}\psi_{12}}{1.0\mathrm{x}1.0 - \mathrm{Cov}(\eta_1\eta_2)\mathrm{Cov}(\eta_1\eta_2)} . \qquad 6.5$$

This implies that, other things being equal, the coefficient of determination would encourage us to prefer models with larger error covariances.[6] Is this justifiable? This may be acceptable in the extreme. A covariance implying a correlation of 1.0 between the errors might say that there really is only one error variable, and hence that this model is simpler because only a single error variable is operative in influencing both the η's. But in general, this deepens rather than reduces the mystery. Here we are being encouraged to value a common error as a cause, whereas the discussion above convinced us that common causal factors were to be devalued! Are we really to value common causes when we can't identify them and to devalue them when we have identified them as a common factor?

We might try to salvage some meaning for the coefficient of determination by considering three dependent variables but this seems to make matters worse, not better.[7] Overall, I am led to conclude that we should ignore the LISREL manual's advice to use the coefficient of determination to assess models until someone explains the utility of using these particular combinations of error and concept variances and covariances.[8] If one is merely looking for an indication of whether the endogenous variables as a set are well or poorly accounted for, the average of individual R^2 would seem sufficient.

We should note that the LISREL 7 manual (Joreskog and Sorbom, 1989:41) defines a parallel coefficient of determination for the observed variables as 1.0 minus the determinate of θ divided by the determinate of **S** (presumably meaning the portion of **S** corresponding to Θ_ϵ or Θ_δ). This calculation suffers from the same difficulties noted for the coefficient of determination, and hence seems destined to land in the same dustbin.

6.2 Seeking a Motif Model

More than a few of our statistician friends have used Monte Carlo studies as a magnifying glass to look for dirt and crawly things in the corners of our statistical house.[9] Monte Carlo investigations of the behavior of various estimators, test statistics, and fit indices have turned up some embarrassments but considerable uncertainty remains because it is unclear whether there really are dirt piles in our statistical corners or whether flecks of factor dust are obscuring the Monte Carlo magnifying glass. More often than not, the factor model serves as the test model in Monte Carlo investigations, and hence the factor model confounds the ultimate decision of statistically clean or unclean.

The problem is that the factor model is no longer typical of the models being estimated, and hence demonstrating the deficiencies or strengths of particular statistics in the context of the factor model is no longer a convincing demonstration that those same deficiencies or strengths will be found in the corners of research reality. Monte Carlo demonstrations would be more convincing if the superior/inferior behavior of a particular test or estimator were established using a test model corresponding more closely to models currently in use. A Monte Carlo verdict is convincing only if the test model encapsulates the gist or motif of current modeling practice.

No single test model could possibly reflect all the characteristics of all current models, but this is no excuse for failing to develop a test model that comes closer to capturing the motif of current practice. Without a standard or motif model, Monte Carlo devotees will be forced to use ad hoc test models, which hinders comparability across studies,[10] or to use some variant of the factor model, which dusts the Monte Carlo magnifying glass with factor flecks, leaving us to guess whether our nonfactor models are statistically clean or dirty. The following is proffered in the hope of initiating discussion about the specification of a more typical, or motif, test model.

The characteristics I would like to see represented in a motif model include a reciprocal and a looped effect, concepts with one, two and three indicators, a shared indicator, a fixed zero exogenous covariance, indirect effects, and error covariances at the measurement and structural levels, all embodied in a moderately parsimonious structural model. Figure 6.1 integrates these components into a relatively small model that is intentionally congested near η_6 to emulate the unevenness of most models and to provide a focal point for likely estimation difficulties.[11]

The motif model in Figure 6.1 is a bit atypical of current practice in that both a loop and a reciprocal effect are included. The discussion in Chapter 3

Figure 6.1 A Motif Structural Equation Model.

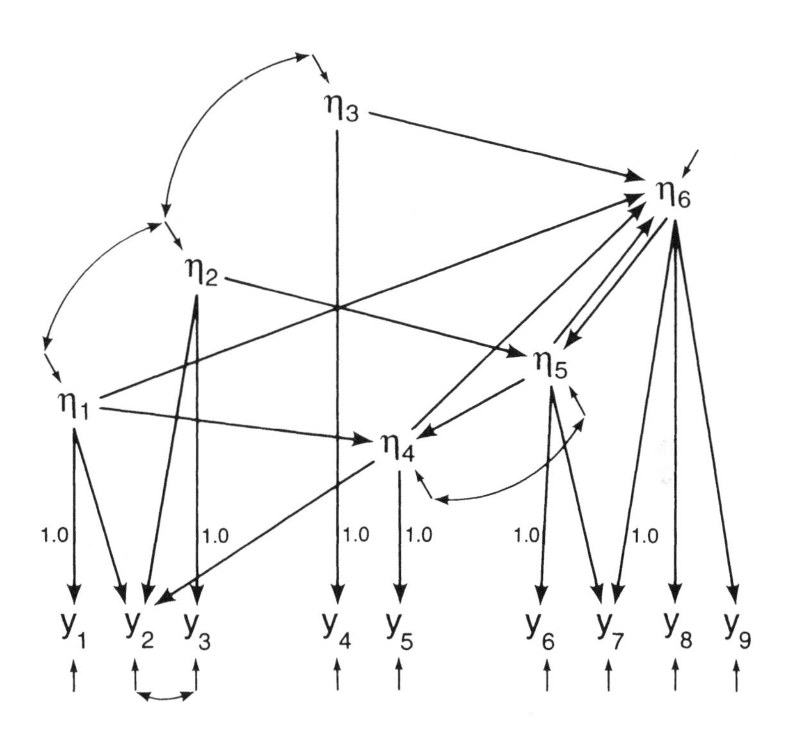

and the ubiquity of discussions of feedback and mutual responsiveness convince me that structural equation modeling will steadily progress towards use of more loops. Consequently, I have included the reciprocal and larger loop (η_5 to η_4 to η_6 to η_5) with the intent of encouraging investigation of a relatively complex model feature that is likely to be fundamental in the future.

This motif model does not include any indicatorless concepts, error-free concepts arising from definitional relationships, fixed non-zero effects, indicators which cause concepts, or equality constraints.[12] I eliminated these from the motif model because I do not think these styles of model segments are currently used, or are likely to be used, with great frequency. As a set, however, these characteristics are not infrequent. So perhaps in addition to the motif model, Monte Carlo studies might also examine a nuance model that incorporates more

Figure 6.2 A Nuance Structural Equation Model.

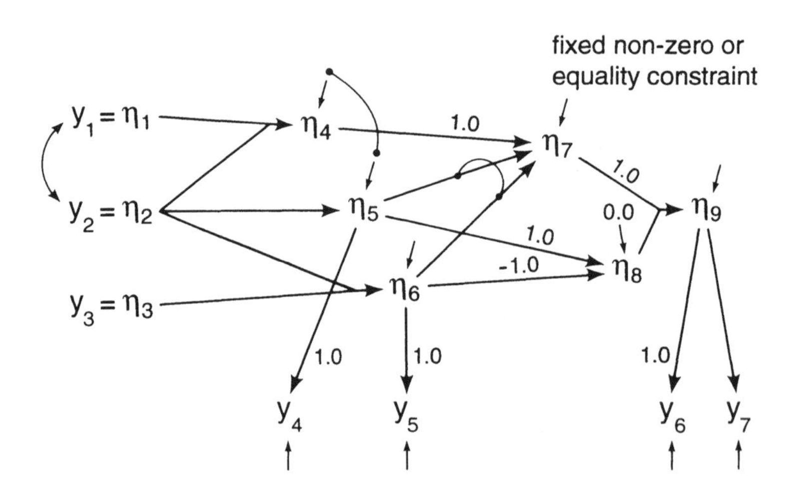

subtle distinctions and characteristics. Figure 6.2 proposes a possible nuance model which includes two concepts with no indicators, one of which (η_8) is determined by definitional connections to other concepts,[13] the other of which (η_7) is scaled through both its assumed but unestimated dependence on η_4 and its effect on η_9. The concept η_4 has two causal indicators (y_1 and y_2) and η_9 has two effect indicators. And two concepts in the nuance model (η_5 and η_6) have both causal and effect indicators. Several fixed non-zero coefficients are included and there are two equality constraints (depicted with curved lines terminating in dots). One constraint equates two variances while the other equates two effects.[14]

I hope that both the factor model and whatever standard motif model is eventually adopted will be incorporated into Monte Carlo studies until the differences between the behaviors of these models are relatively well established. I view the nuance model as being appropriate for a more limited set of Monte Carlo concerns, and hence as being likely to be used less frequently. Regrettably, I must end by repeating the caution that dependence on the traditional factor model renders many Monte Carlo assessments of statistical cleanliness or uncleanliness untrustworthy in direct proportion to the degree to which one's modeling style has progressed beyond postulation of correlated common causes.

6.3 Links to the Literature

My final housekeeping task is to organize the literature that has been accumulating while we were occupied with other matters.

6.3.1 Goodness of Fit Indices

Numerous recent articles have been devoted to creating, evaluating and critiquing indices of model fit.[15] The November 1992 issue of *Sociological Methods and Research* was devoted to tests appropriate for structural equation models, with articles by Bentler and Chou (1992), Bollen and Long (1992), Bollen and Stine (1992), Browne and Cudeck (1992), Gerbing and Anderson (1992), and Long and Trivedi (1992). These articles, along with additional works by Joreskog (1993), Muthen (1993), Raftery (1993), Saris and Satorra (1993), Tanaka (1993), and Wothke (1993) were published as a book on model testing edited by Bollen and Long (1993). Other discussions and/or definitions of fit indices appear in Bentler (1990a,b), Bollen (1989a,b), Marsh and Balla (1994), and Joreskog and Sorbom (1993:125).

The quantity of material appearing on this topic stands in stark contrast to the thinness of the current consensus on how to proceed. Bollen and Long (1993:6-7) summarize the points of consensus as:

1) Strong substantive theory is fundamental to assessing model fit. "If the model makes little substantive sense, it is difficult to justify." (Bollen and Long, 1993:6).

2) The χ^2 test should not be the sole basis for determining model fit because a variety of concerns may lead to rejection of worthwhile models. Specifically, excessive test power (due to large N)[16] may prompt the rejection of acceptable models; failure to satisfy the multinormality assumption works against finding acceptable fit; and the χ^2 test makes no allowance for the approximate, loose, rough, or simplifying nature of social science models.[17]

3) "No single measure of overall fit should be relied on exclusively ... there is little benefit to turning from a reliance on the χ^2 test statistic to a reliance on any other single fit index" (Bollen and Long, 1993:6).

4) Even models with excellent fit indices may be unacceptable because

other components of the model may be unacceptable. Wrongly signed effects, negligible R^2, or signs of improper solutions are not negated by a high value on any overall fit index.

And finally, 5) specify and examine the relative performance of alternative models.

These points reiterate themes that have been viewed as desirable operating procedure for some time, and hence should prompt no change whatever in the modeling habits of those who are already conscientious modelers.

I am surprised that Bollen and Long did not detect a consistent concern for using patterns in the residual matrix as a sign of ill fit. After all, it is the difference between the **S** and **Σ** matrices that drives all tests of overall fit, and systematic differences here, even if small, warrant caution. It is precisely the inability of any single measure to reflect the innumerable possible residual patterns that might be of substantive importance that constitutes the major hinderance to developing a universal fit statistic.

The variety of research objectives that can prompt model construction also contributes to the elusiveness of any generally applicable fit index. Research goals range from purely exploratory modeling attempts, to seeking minor modifications on already established models, to seeking alternative models, to seeking patterns among sets of coefficients, to examinations of alternative indicator sets, and so on. It is unreasonable to think that a single overall fit criterion would be equally applicable to such diverse modeling objectives.

In some contexts one may want a comparative measure of fit, so that a model is assessed relative to some other substantive model rather than to an artificial null model. In other contexts one may wish to reward models for their parsimonious use of coefficients while penalizing models that are saturated with coefficients.

The Akaike information criterion (*AIC*), for example, is an index that rewards sparse models, or penalizes more heavily parameterized models, but even this relatively well-known coefficient is not without its problems. One set of problems is created because Akaike (1987:321) provides two different definitions for the *AIC*. The first of Akaike's definitions follows from his adopting a specific conceptualization of "information" (Kullback-Leibler information), and then relating this to maximization of the log likelihood, which results in the definition of *AIC* as

$$\mathrm{AIC} = -2\log L(H) + 2t \qquad 6.6$$

where *logL(H)* refers to the maximum of the log likelihood attainable by adjusting the estimates of the *t* unknown parameters comprising the hypothesized model *H*. Given that smaller *AIC* values are desirable, it is clear that the penalty the *AIC* attaches is twice the number of estimated parameters comprising any given model. Equivalently, the AIC might be said to reward parsimony by twice the number of omitted parameters. But what does a reward or penalty of two units per parameter estimate amount to? Is this a strong penalty or a weak penalty?

Akaike's second definition arises when he attempts to provide another perspective on the nature of the parsimony reward/penalty by simplifying the above formula. Akaike uses the above formula to obtain an *AIC* value for the model of interest *H* and another *AIC* value for a saturated model H_s having *s* coefficients, which we designate as *AIC(H)* and $AIC(H_s)$ respectively. The difference between these two quantities is

$$\mathrm{AIC(H)} - \mathrm{AIC(H_s)} = -2\log L(\mathrm{H}) - (-2)\log L(\mathrm{H_s}) - 2s + 2t. \qquad 6.7$$

The right side of this equation contains a difference between two logs, which is the same as the log of a ratio,[18] and which in turn leads to a variable having a χ^2 distribution because minus twice the log of a likelihood ratio has a χ^2 distribution. The degrees of freedom for the χ^2 equal the difference between the number of parameter estimates required in the saturated and unsaturated models, namely $df = s - t$, and these same degrees of freedom can be used to express the difference between *s* and *t* on the right side of the equation as follows:

$$\mathrm{AIC(H)} - \mathrm{AIC(H_s)} = -2\log(L(\mathrm{H})/L(\mathrm{H_s})) - 2(s - t) \qquad 6.8$$

$$\mathrm{AIC(H)} - \mathrm{AIC(H_s)} = \chi^2_{df} - 2df. \qquad 6.9$$

The second definition of *AIC* arises when Akaike (1987:321) states that by "neglecting the common constant" $AIC(H_s)$ "we may define the *AIC*" for the model *H* as

$$\mathrm{AIC} = \chi^2_{df} - 2df. \qquad 6.10$$

$AIC(H_s)$ is a constant because the saturated model will never change though it has some likelihood and a non-zero number of parameters as required by definition one above. *AIC(H)* for the hypothesized model is not a constant, as this depends on which particular nonsaturated model is considered.

In this form it is clear that the penalty the *AIC* applies for non-parsimony in the number of estimated parameters can be calibrated in χ^2 units. Specifically, a parsimony penalty/reward of two χ^2 units is assessed for each additional estimated/nonestimated parameter.

This is a moderate to heavy[19] penalty, but we have a more serious problem. We now have two definitions of *AIC* (Eq. 6.6 and Eq. 6.10). Some authors use the first (Bandalos, 1993), others use the second (Joreskog, 1993), others use both (Kaplan, 1991), and still others seem to mix the two of these, as when Joreskog and Sorbom (1993:119) indicate that LISREL's calculation of the *AIC* follows the form of the first definition but verbally describe the likelihood term as a χ^2. We are left to guess which of the two definitions has been implemented in LISREL, or indeed whether Joreskog and Sorbom (1993:119) have implemented a third definition of *AIC* because their verbal description might also mean that they eliminated *2s* (another constant) from the right of the second definition.[20]

The elimination of constants, in moving from definition one to definition two by Akaike (Eq. 6.6 to 6.10), or from definition two to definition three by Joreskog and Sorbom (like Eq. 6.10 but with *-2t* instead of *-2df*) will not alter the comparative decisions one arrives at if one has a single data set and several nested alternative models. From the perspective of any one of these *AIC*'s, the researcher should favor the model with the smallest *AIC*.

But equally clearly, the now-you-see-it-now-you-don't messing with the constants implies that the *AIC* is completely unnormed, and hence no general guidelines can be provided for how big an *AIC* value is unacceptable or how small a value would be fantastic. The only utility the *AIC* has is as a comparative index linking models based on the same data set, and consistently using any one of the definitions. Adding or deleting even a single indicator of any one concept in a model will change the log likelihood and hence will change the *AIC* in any of its forms. This makes comparative use of the *AIC* with even overlapping data matrices questionable, and effectively eliminates the current versions of the *AIC* as contenders for a general fit index.

Another fundamental question arises because there is some lack of agreement as to whether the parsimony penalty applied by the *AIC* is the most appropriate penalty. This issue was raised indirectly when Bozdogan noticed that the *AIC* was not a consistent estimator, and hence sought to "improve and extend" it (Bozdogan, 1987:357). Bozdogan (1987:358) proposed a consistent *AIC* estimator (*CAIC*) which is calculated as

$$\text{CAIC} = -2\text{LogL(H)} + t(\log(n) + 1) \qquad 6.11$$

where t is still the total number of estimated coefficients and $log(n)$ is the natural logarithm of the sample size. The difference in the parsimony reward/penalty applied by this formula becomes clearer if we express this as

$$\text{CAIC} = -2\text{LogL(H)} + t + t\log(n). \qquad 6.12$$

For a sample size whose logarithm is 1.0, the parsimony correction would be *2t*, and the *CAIC* would be identical to Akaike's first definition of the *AIC*. Since the natural logarithm of 2.71 is 1.0, this implies that it takes a very small sample (a sample size of about 3) for the *AIC* and *CAIC* to apply similar parsimony penalties. For larger sample sizes the *CAIC* applies a larger penalty. In particular, for sample sizes of 150 and 1000 the *CAIC* applies penalties of about *6t* and *8t* respectively, or penalties about three to four times the *2t* penalty assessed by Akaike.

Bozdogan warns that "the virtue of consistency should not be exaggerated" (1987:356), thereby implicitly critiquing his own index, but this does not seem to have deterred the propagation of the *CAIC*, which is now reported by both EQS (Bentler, 1992a:92) and LISREL (Joreskog and Sorbom, 1993:119) immediately after the *AIC*. The user is left to decide which to use, though no basis on which to decide is currently available. All one can do is hope that the *AIC* and *CAIC* give the same ranking of the competing models so that the choice is moot.

But even this does not exhaust the complicating details. Depending on the perspective with which one approaches the *AIC*, one can be led to conclude that the "choice among the modified competing models should be made via the *AIC*" (Kaplan, 1991:307)[21] or "that Akaike's information criterion cannot be used for model selection in real applications" (McDonald and Marsh, 1990:247). In this context it is not surprising that the authors of the main structural equation programs (Joreskog and Sorbom, 1993 and Bentler, 1992a) seem unable to distil the indices they report into specific followable advice.

Lacking a clear notion of precisely what it is that is to be summarized about a model by any fit index, and lacking any agreement on the characteristics that such an index should have,[22] the current trend seems to be to define measures that cover the full range of possibilities, so that one has a measure for each and every combination of the potentially desirable characteristics. Ultimately, the lack of agreement on the desirability of specific fit index characteristics comes back to haunt practicing researchers because the availability of even hundreds of indices will be neutralized by the absence of a consensus on which specific combination of characteristics is the proper combination for use in a particular context.

Researchers interested in structural equation modeling as a tool, and not as a vocation, are advised to avoid detailed pursuit of the plethora of new fit indices for the next few years. *For most purposes it is sufficient to report the* χ^2 *test (and its probability) as long as this is complemented by a discussion of model parsimony as evidenced in the* χ^2 *degrees of freedom, a discussion of the degree of test sensitivity provided by the sample size, a report of the adjusted goodness of fit index (AGFI), and a discussion of any noticeable patternings in the residuals.*[23] Having fully reported and discussed χ^2 and having mentioned the AGFI index, one should turn one's attention to more detailed indices of particular components of model fit,[24] or to alternative models, or to equivalent models, or to considerations of the styles of data that might substantively distinguish between models, or whatever it is that is of import in the relevant literature. Only a nearly fanatical dedication to wanting to know "the" fit of a model should prompt consideration of even half the fit indices listed in the references above.[25]

6.3.1.1 Fit, the Modification Index, and Expected Parameter Change

Discussions of model fit are complemented by discussions of model modification, or specification searches, because most searches and modifications are designed to improve a model's fit. We have already examined model modification from the perspective of TETRAD in Chapter 4, so we are free to examine other aspects of the current literature. It is gratifying to see that the literature has reached a consensus on two points, namely *that models should be only minimally modified and that each and every modification should be reported.*[26] Together, these minimize capitalization on chance and maximize the attention paid to the theoretical implications of the modifications.

There has also been some progress on the development of new statistics to assist in model modification. Kaplan (1989b, 1990) recommends use of an expected parameter change (*EPC*) statistic as an additional piece of evidence to examine during model modification, and LISREL (Joreskog and Sorbom 1993:149) now provides both a basic and standardized version of this statistic.[27]

There is a close conceptual connection between the modification index and the new expected parameter change statistic. Consider Figure 6.3, which is a hypothetical depiction of a fit function that is to be maximized through selection of an optimal estimated value for a model's coefficient. In this instance the values of the fit function are scaled in units of χ^2 so that the best fit is attained at the minimum value of the fit function. The solid curve indicates the best fit attainable for a variety of possible values for one of the model's coefficients, namely a β that is currently not in the model because its value is fixed at zero.[28] Because the current fixed value of the β coefficient is zero, we are imagining that the model we are attempting to improve has the χ^2 value highlighted by a dot on the vertical axis.[29] The solid curved line represents the fit function and indicates that if this β coefficient were freed, the fit could be improved by increasing the estimate for β but that too large an increase in the β estimate would result in a deterioration in fit.

If the β coefficient's value was freed, successive iterations would follow the fit function down toward the dot at the minimum of the fit function. This dot reports both the specific estimate of the β coefficient that would arise if this coefficient was estimated (the β value found directly below the dot), and the best fit attainable by freeing the β coefficient (the χ^2 fit appearing directly to the left of the dot at the minimum of the fit function).

For any particular data set and model, the exact shape and curvature of the fit function is unknown unless it is actually traced out by trying different values for the coefficient, but the fit function can be closely approximated, as

Figure 6.3 Expected Parameter Change.

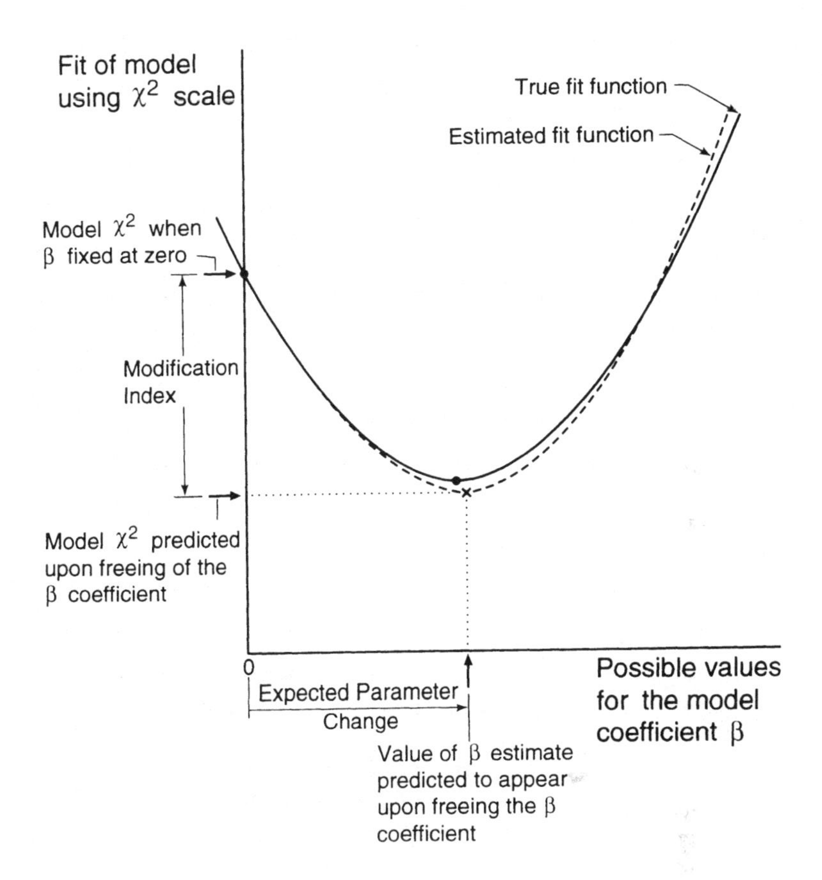

indicated by the broken curved line. The minimum of this approximate fit function (denoted by an x) provides an approximation of the best attainable χ^2 fit, and an approximation of the value of the β coefficient that would provide that fit. These approximate estimates provide the modification index and expected parameter change statistics for this β coefficient. Specifically, the modification index is the difference between the current and predicted model χ^2's as indicated on the vertical axis, and the expected parameter change statistic is the difference between the current (zero) and predicted estimate of the β

coefficient, as indicated on the horizontal axis.[30]

Naturally, statisticians will attempt to provide an ever-improving match between the true and approximated fit function because the better the approximation of the broken curved line to the true fit function (the solid curved line) the closer the modification index and expected parameter change statistics will be to the actual changes in χ^2 and coefficient estimate that appear when one frees the coefficient and locates the heavy dot at the true minimum of the fit function.[31]

The logic behind using the modification index (predicted change in model χ^2) is straightforward and is relatively widely known. Freeing a coefficient with a large modification index will provide a large improvement in the fit of the model, with all the attendant temptations and limitations.[32] The issue raised by Kaplan is whether there is any utility in considering the size of the anticipated change in the magnitude of the coefficient estimate (the *EPC*) as one tries to decide whether or not to free the β coefficient.

In most instances, researchers will turn to the *EPC* statistic only after they have noticed that the modification index suggests that freeing a particular coefficient would lead to a significant improvement in model fit.[33] Few researchers will choose to, or not to, free a coefficient on the basis of the specific magnitude of the coefficient that is likely to result, so the primary use of the expected parameter change statistic is likely to hinge on its ability to report the sign of the anticipated change in the coefficient's value. If the anticipated sign is unreasonable, the large modification index should be ignored because it is reporting that the improvement in model fit could only be attained by inserting an unreasonable coefficient value into the model. If the modification index is large, and the sign of the *EPC* is reasonable, then the coefficient may be freed.[34]

Note that the expected parameter change statistic is scaled in whatever real units the corresponding model coefficient possesses. In the case depicted in Figure 6.3 the expected parameter change would provide an estimate corresponding to an unstandardized β slope (like a b in multiple regression). Just as multiplying an unstandardized effect coefficient by the standard deviation of the causal variable and dividing by the standard deviation of the dependent variable results in a standardized slope,[35] multiplying and dividing the expected slope (the *EPC*) by the same standard deviations provides a standardized version of the expected parameter change (*SEPC*). Similar adjustments for the metrics of relevant variables standardize expected parameter changes corresponding to covariances into expected correlation changes, and metric error variances into proportions of error variance.[36]

Regarding model modification more generally, one should not lose sight of the fact that only a few lucky researchers are likely to encounter failing models that can be brought to authenticity by the freeing of a few reasonable coefficient estimates. Most failing models will probably require wholesale model revisions, insertions of entirely new concepts, or reorganizations of the whole model. The researcher with a failing model is more likely to have to develop a whole new conceptual framework, as opposed to merely tinkering with the existing framework (Kaplan, 1989b). It is these deeper and more challenging searches that are likely to produce the greatest benefits. As Hayduk (1990:196) puts it, if structural equation models are "to be prods to sluggish imaginations, sparks that ignite insight, keys that unlock advancement, or hammers that forge progress from burning issues, we will have to do better than merely searching through the list of potentially-freeable coefficients, no matter how diligently and with how much technical sophistication we conduct the search."[37]

6.3.1.2 Fit, Other Indices, and Power

There has been some discussion of the Lagrange Multiplier and Wald tests that are available in EQS but not LISREL. The Lagrange Multiplier test is much like LISREL's modification index in that it attempts to predict the reduction in the model's χ^2 that would accompany the release of an equality constraint or the freeing of a currently fixed parameter (Bentler, 1992a:68; Chou and Bentler, 1990:117). It differs from the modification index in that it can also be used to test for the simultaneous freeing of several currently fixed or constrained coefficients. The Wald test, in contrast, is concerned with eliminating a set of one or more unnecessary coefficients from a model (Bentler, 1992a:68).[38]

The Lagrange and Wald tests are asymptotically equivalent to L^2 or the usual χ^2-difference test (Bentler and Chou, 1992; Chou and Bentler, 1990:117; Raftery, 1993:167; Satorra, 1989), so comparable tests are available to LISREL users. If several parameters are to be considered for simultaneous freeing or fixing, the LISREL user will pay the price of having to estimate two models, one with the set free and the other with the set fixed, but in addition to obtaining the desired test statistics they also gain the benefit of two entire sets of diagnostics and the new set of coefficient estimates. The EQS user of the Wald and Lagrange tests who concludes that a modified model is acceptable will also be faced with doing a second model estimation to obtain estimates for the new, and presumably preferable model. There are instances when the χ^2-difference test seems to outperform the Wald and Lagrange tests (Chou and Bentler, 1990, and see also Kaplan and Wenger, 1993), so a conservative approach would seem to dictate continued use of the χ^2-difference procedure until such time as the Wald and Lagrange tests offer some specific superiority.

Bentler and Chou (1993) consider whether any diagnostic potential can be wrung from changes in the estimates and sampling variability of already free coefficients that accompany the freeing of an additional coefficient. This quest represents a substantial departure from the style of information provided by the modification indices and expected parameter change statistics, but this too is not yet worthy of a blanket endorsement. Indeed, "substantial additional experience will be needed before unambiguous recommendations regarding the role of these statistics in model modification can be made" (Bentler and Chou, 1993:236).

The discussions of fit indices and coefficient diagnostics have been complemented by discussions of power (Joreskog and Sorbom, 1988:217-218, 1993; Kaplan 1989c; Kaplan and Wenger, 1993; Matsueda and Bielby, 1986; Saris and Hartman, 1990). Practical, though not hand-calculable, procedures for determining the power of model tests against nested alternative models are becoming available (McArdle, 1994; Saris and Satorra, 1993), but it is still unclear as to when these will become sufficiently user-friendly to be recommended as a required part of routine practice.

Another perspective on model fit can be found in the literature on cross-validation. Cross-validation refers to estimating a model on a calibration sample and then checking that model on a validating sample. Cross-validation has been of little interest because most researchers do not have the luxury of two samples, one for model estimation/development and the other for validation. More interest in cross-validation is likely to be shown in the future because the need for two samples has evaporated. Browne and Cudeck (1989) discuss a procedure for estimating a cross-validation index from a single sample. Several other authors are also working on cross-validation (Bandalos, 1993; Camstra and Boomsma, 1992; MacCallum et al., 1994; Opwis, Gold and Schneider, 1987) so we are likely to be treated to discussions of the similarities and differences between any or all the fit indices and the cross-validation index in the not too distant future. Bandalos (1993:353), for example, provides a cross-validation interpretation of the *AIC* coefficient.

I am inclined to end this quick overview of fit precisely where Browne and Cudeck (1993:157) end, namely with an admonition that "fit indices should not be regarded as measures of usefulness of a model." Browne and Cudeck had in mind that model fit and model plausibility are very distinct concerns. To this I would add that model fit and model utility are equally distinct. Discussions of failing models can be of great utility, as was argued in Chapter 1.

6.3.2 Are Effects Causal Effects?

Sobel (1990) discusses the interpretation of causal effects in structural models and concludes that some current interpretation practices are unfounded. Sobel sees interpretation problems as arising because researchers do not routinely distinguish between "factors" ("variables the researcher can manipulate at least in theory," 1990:499) and "concomitants" ("variables that it does not make sense to manipulate" even "hypothetically," 1990:499). Sobel suggests that variables that are mere concomitants should not contribute to the calculation or interpretation of any effects. The mental and statistical bookkeeping required to keep track of the segments of models with permissible and forbidden causal interpretations seems to Sobel to pose a major unrecognized challenge, and hence Sobel proffers some cautions regarding acceptable interpretations.

But be warned, Sobel's conclusions "hinge on the use of (a particular) perspective" (1990:498, parenthetic material added), and the perspective is unlikely to receive general endorsement. Space does not permit a full examination of Sobel's arguments, but we should point out that one of Sobel's key tenets is at odds with the perspective on measurement taken in Chapter 1. Hence, if you are persuaded by my perspective on measurement, you too will find Sobel's arguments unconvincing.

To get a peek into this particular messy corner consider that one can adopt a perspective that seems to make Sobel's definitions of a concomitant variable a contradiction. A variable must have some variance, or else it is a constant. If a variable has variance, it must make sense that something or someone can change or manipulate the values that provided that variance. Hence, it is contradictory to define a concomitant as a "variable" (something whose variance makes sense as something's or someone's manipulations) "that it does not make sense to manipulate" even hypothetically (Sobel, 1990:499).

The connection of this to the measurement discussion in Chapter 1 becomes clearer when Sobel (1990:504) provides a discussion of IQ as an example of what it would take to turn a concomitant into a factor, and hence to render standard causal interpretations for a variable acceptable. In brief, Sobel claims that what it takes is recognition that individuals' true scores on the concept (IQ or ability) may differ from their scores on the corresponding indicator. That is, if one has a sufficiently precise conceptualization of a concept to permit recognition of things that interfere in the measurement of that concept, then one also has sufficient ability to make sense of the variance in that concept to call the variable a factor, instead of a troublesome concomitant. That is, researchers capable of conceiving of their concepts as distinct from their indicators need not be concerned with the interpretational difficulties associated with concomitants because the recognition of measurement error simultaneously acknowledges sufficient meaningful and sensible variation in the concepts to keep them from becoming concomitants. In short, if you function in accordance with Section 1.5, you are saved from most of the interpretational difficulties posed by Sobel (1990).[39]

6.3.3 Multi-trait-multi-method Analyses.

Multi-trait-multi-method (MTMM) analyses have also received a lot of scrutiny. Frequent estimation problems (Marsh, 1989) prompted examination of the identification of MTMM models, and resulted in the location of some underidentified and nearly underidentified MTMM models (Kenny and Kashy, 1992; Grayson and Marsh, 1994). Marsh and Byrne (1993) suggest that constraints between groups can be used to make MTMM models more parsimonious, though I know of no demonstration of how parsimony created with equality constraints between groups improves the identification of the problematic segments of MTMM models.[40]

Byrne and Goffin (1993) examined four alternative approaches to analyzing MTMM models, and recommend that since no single approach is clearly superior, researchers should estimate a variety of MTMM models and use the best fitting model. Henley (1993) examined the robustness of some estimators in the context of MTMM models, and Levin suggests a three-step approach to MTMM estimation that "transforms the solution to a classical multiple group factor analysis" (1988:469).

It is true that traditional MTMM thinking has a strong factor analysis slant. In the MTMM context, it is taken as given that one is seeking methods factors and trait factors, both of which function as common causes of specific items. MTMM thinking has moved beyond traditional factor analysis to the extent that specific factors are postulated as leading to specific items. Each item is typically grouped under one method and one trait but the main thought style is still to rely on indicator covariance as resulting from dependence on common factor causes.[41]

Some of the prime sources of differentiation among the various MTMM models concern whether some of the loadings are to be equal, whether error covariances are permitted, and whether the factor covariances are free. From the perspective of Chapter 1, this makes it seem as if the users of MTMM are being pushed toward structural equation modeling by a glacier. It is precisely when one makes the move toward permitting, indeed toward seeking, effects (not correlations) among the factors that one moves away from traditional factor thinking, and it is precisely this move that provides the greatest hope for improving the identification status of any MTMM structural model.

The MTMM literature seems to be seeking a model that is appropriate whenever one has measured several suspected traits in different ways, just as the factor model was supposed to be the appropriate procedure to locate the traits underlying several indicators. Attempting to measure several traits with several measurement styles does not guarantee that one particular style of trait and method model will necessarily constitute the appropriate model.[42] Indeed, given the variety of research contexts in which multiple traits and methods are employed, the search for a standard MTMM model seems like a search for a cage.

The prolonged search for a standard MTMM model can only distract researchers from the alternatives that already exist. The examination of external validity (e.g., Marsh, 1989) will eventually exterminate any standard MTMM model because the specification of external benchmarks will require nonstandard models. But perhaps the most efficient way to rid ourselves of this particular mess would be to stress the need for alternative models by requesting that at least one nonfactor model be specified as an alternative to the author's favored MTMM model. Without examination of at least one substantively different model, the author's claims to having assessed his or her model are suspect. There really is no substantive alternative model if the only "alternatives" are minor variations on the multiple traits and methods theme. What reasons can the champions of MTMM models provide to justify why their particular corner of the structural equation house should be spared the cleansing provided by the demand for substantively alternative models?

6.3.4 Multilevel data

One of the most promising developments in the literature is the appearance of analysis strategies appropriate for multilevel data, such as the clustering of individuals within families, individual within groups, or children within classrooms within schools (DiPrete and Forristal, 1994; Ellis, 1993; Goldstein and McDonald, 1988; Harnqvist et al., 1994; Lee and Poon, 1992; McDonald, 1994; McDonald and Goldstein, 1989; Muthen, 1989a,b, 1994; Willett and Sayer, 1994). Standard programs such as LISREL and EQS can be used to estimate multilevel models (Lee and Poon, 1992; Muthen, 1994), though McDonald's (1994) BIRAM (bi-level reticular action model) may have a slight long-term advantage due to its use of a minimal number of partitioned matrices.[43] Unfortunately, the factor model seems to have again garnered center stage in these developments, and this may delay the investigation and propagation of procedures appropriate for investigation of non-factor models.

Muthen's (1994) procedure for estimating multilevel models within LISREL uses LISREL's multiple-group capabilities to estimate the model, but the groups are no longer parallel groups, they are imbedded groups. This imbedding implies that different predictor and explanatory variables are likely to be appropriate in the two model segments, which in turn will increase the need for the simultaneous estimation of data matrices based on differing sets of variables, as discussed in Chapter 5. Muthen (1994) and McDonald (1994) provide helpful introductions to multilevel models, but as usual, more than a minimal number of pitfalls await the incautious.

In this instance, I think the payoff would be well worth the academic effort required to establish a new academic order. Many of the pitfalls concern interpretation, and understanding what is and is not explained as one moves between levels. Additional precision in these aspects of multilevel structural models should prod us into sharpening our thinking about the nature of reductionistic arguments. I suspect the relatively stale reductionistic theoretical corner is about to receive a good statistical airing-out! Multiple-level models constitute my guess as to the segment of the structural equation modeling literature where additional reading should have the greatest long-term payoff for practicing researchers.

6.3.5 On Partitioning Error Variables

There have been several discussions of the partitioning of error variables into stable and unstable error components in longitudinal designs (see Matsueda, 1989; Palmquist and Green, 1992; Raffalovich and Bohrnstedt, 1987; and Zeller, 1987). Once prompted to think of both stable and unstable components of error variables in longitudinal models, it seems natural to consider whether error variables in nonlongitudinal models are similarly partitionable.

Figure 6.4 depicts how one might model an error variable as "segmentally independent," by which I mean that part of the error variable is independent and part is not. In Figure 6.4 the error attached to η_5 has been replaced by the error variable η_n which is segmented in that it arises from both a statistically independent variable $\xi_{\#+1}$ and a variable $\xi_{\#}$ which is correlated with one of the exogenous variables. The variances of $\xi_{\#+1}$ and $\xi_{\#}$ are constrained to be equal, and the fixed effects of these ξ's on η_n permit assertion of differential contributions of the dependent and independent components to the error variable η_n.[44]

A model specified as in Figure 6.4 but without the $\xi_{\#}$ covariance should provide estimates identical to those obtained with a usual independent error model specification because the attached model segment has only one coefficient to be estimated, and this estimate mimics the contribution made by the original estimate of η_5's error. Such a model would have the additional advantage, however, of including modification indices for the covariances between $\xi_{\#}$ and the other model ξ's, and hence would provide diagnostics indicating which of these covariances might be identified and substantial.[45] Since error variables make strong contributions in most models, the specification of segmentally independent error variables has the possibility of making substantial contributions to our understanding of model functioning.

Figure 6.4 Partitioning an Error Variable.

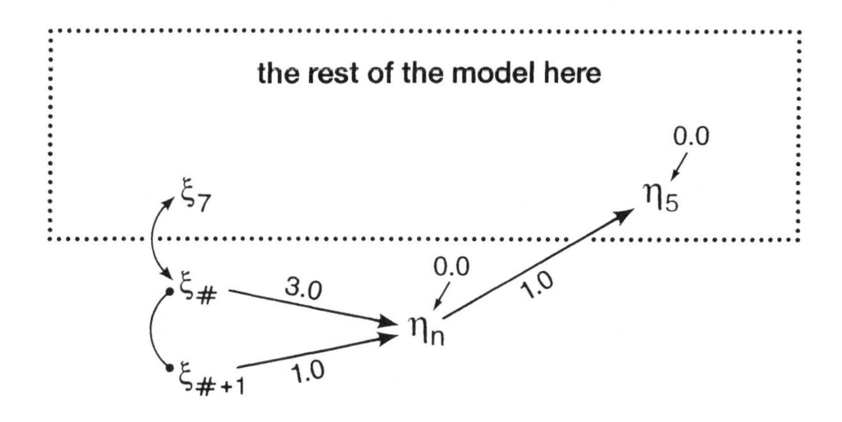

6.3.6 Tying Up The Final Loose Ends

At the end of any cleanup, there are always odds and ends which are too good to discard, yet which have no obvious home. We follow the well-established housekeeping tradition of tying these into bundles and stashing them away until someone needs them.

If you ever need more on **panel or longitudinal models** see Gollob and Reichardt (1985, 1987), Greenberg and Kessler (1982), Joreskog and Sorbom (1980), Marsh (1993), Matsueda (1989), Mayer and Carroll (1987, 1988), Oud, van den Bercken and Essers (1990), Palmquist and Green (1992), Raffalovich and Bohrnstedt (1987), Westholm (1987), and Willett and Sayer (1994). The combination of cross-sectional and longitudinal data is discussed and illustrated by McArdle et al. (1991).

For a procedure for massaging the data so that **dynamic factor models** can be estimated in LISREL see Wood and Brown (1994). For modeling change over time see Willett and Sayer (1994) and for other advances on the basic factor model see McArdle (1994) and McArdle and Cattell (1994).

On **ipsative measures**[46] see Chan and Bentler (1993) and Dunlap and Cornwell (1994).

For continuing discussion of **means in structural models** see Byrne, Shavelson and Muthen (1989), Dolan and Molenaar (1994), Kuhnel (1988),

Millsap and Everson (1991), Ogasawara (1990), Satorra (1992), and *Essentials* (286-322). For examples of actual analyses using means see Baer, Grabb and Johnston (1991), Marsh and Grayson (1990), and Willett and Sayer (1994).

Strong **outliers** in the data may create biased estimates and improper solutions (Bentler, 1992a:228; Berkane and Bentler, 1988; Bollen, 1987a; Bollen and Arminger, 1991; Gallini and Casteel, 1987). Examination of univariate distributions for outliers should be routine, and may eliminate many of the problems. The current lack of a standard strategy for locating, and more importantly for dealing with, outliers implies that this is likely to remain a thorn in many a researcher's side.

For discussions of **missing data** see McArdle (1994), Mislevy (1993), and Muthen, Kaplan and Hollis (1987).

On **measurement error** see Jaccard and Wan (1995) and Westholm (1987).

Wothke (1993) provides guidelines to understanding how and why the error message of a **"Not Positive Definite"** matrix may arise, and potential solutions to the problem.

On **causal, rather than effect, indicators** see Bollen and Lennox (1991) and MacCallum and Browne (1993), but be warned that about half the problems seen by MacCallum and Browne, and some of their solutions, indicate entrenched factor analytic thinking, as opposed to a dedication to making the model reflect the model the researcher really has in mind.

On more purely **statistical topics** see Arminger and Schoenberg (1989), Benson and Fleishman (1994), Bentler and Dijkstra (1985), Boomsma (1987), Browne and Shapiro (1988), Cudeck (1989), Dijkstra (1990), Dolan and Molenaar (1991), Farley and Reddy (1987), Gerbing and Anderson (1987), Henly (1993), Kaplan (1989a, 1994), Kaplan and Wenger (1993), Lance (1989), McDonald and Hartmann (1992), McDonald, Parker and Ishizuka (1993), Mislevy and Sheehan (1989), Muthen (1989a,b), Muthen and Kaplan (1992), Ogasawara (1990), Satorra (1990), Seidel and Eicheler (1990), Stelzl (1991), Stone and Sobel (1990), and Vittadini (1989).

Browne and DuToit (1992), Cudeck, Klebe and Henly (1993), and DiPrete and Grusky (1990) report on and illustrate procedures for obtaining estimates for a variety of **nonstandard models**. The procedures require user-specified subroutines appropriate for one's model, and are intended to assist those who are sufficiently statistically inclined to be tempted to create a program to suit their specific model. Use of these procedures with the circumplex model are briefly discussed by Cudeck, Klebe and Henly (1993) and Browne and DuToit (1992), and are more thoroughly examined in Browne (1992).

For tricks of the trade that impose **constraints on the variances** of exogenous or endogenous variables see Green and Palmquist (1991) and McArdle and Cattell (1994), and on **inequality constraints** see Kelderman

(1987) and McArdle and Cattell (1994).

For a **Bayesian** approach, see Fornell and Rust (1989) and Lee (1992).

On loops in models recall Chapter 3 and that Chapter 7 of *Essentials* detailed the difference between basic, loop-enhanced and total effects. Sobel (1986, 1987) provides a procedure for determining the significance of specific indirect effects, and these works have been extended and complemented by Bollen (1987b), Bollen and Stine (1990,1993), and Stone and Sobel (1990). Unfortunately, neither LISREL nor EQS provides for testing the significance of user-specified indirect effects. That is, in a complex model, a user should be able to specify a path of interest, say the sequence of paths $\gamma_{21}\beta_{42}\beta_{54}$ which are the path from ξ_1 to η_2 to η_4 to η_5, and have the program calculate the magnitude of this particular indirect effect, the enhanced indirect effect if the path happens to touch a loop, and the significance of this effect. As models become more complex, interest will eventually focus on particular pathways through models, and this will reduce the utility of the significance testing of total and total-indirect effects, as currently provided by LISREL.

The developing understanding of the breadth of the implications of structural equation modeling will eventually lead to a new jargon for discussing **reliability and validity**. This is most easily seen in the context of fixed θ scaling indicators as discussed in Chapter 1, loops in Chapter 3, and concepts without indicators also in Chapter 3, but it is also evident in many more standard discussions, such as Hattie (1985) and Reuterberg and Gustafsson (1992).

It has long been recognized that the use of correlation or covariance data in LISREL becomes suspect if one has ordinal data. **Ordinal variables** have continued to receive considerable attention,[47] though the problems may not be as serious as popular opinion suggests (Homer and O'Brien, 1988; Joreskog and Sorbom, 1988:207; O'Brien and Homer, 1987). Bentler and Chou, for example, suggest that for the near future "continuous methods can be used with little worry when a variable has four or more categories, but with three or fewer categories one should probably consider the use of alternative procedures" (1987:88).

This "better than anticipated" performance of real ordinal data is probably partially due to the fact that in practice Likert and other scales are operationalized as equal-appearing interval scales. While one usually is unable to point to a calibrated dial as proof that one has a continuous and interval scale, standard operating procedures strive for as close an approximation to this as possible. When verbal descriptions of categories are provided they are often conscientiously spaced along the intended continuum precisely to avoid the extremes of clustering possible with mere ordinality.[48]

In adjusting for mere ordinality, PRELIS assumes that the underlying variables are bivariate normal, and that the ordinality of the available response patterns may have resulted in misestimation of the true correlation. Imagine, however, that the true distribution is not multivariate normal. Which is the more serious error, creating estimates out of the nonmultivariate normal distribution directly, or adjusting on the basis of the now-erroneous assumption of multivariate normality and then using the unjustifiably adjusted correlation coefficients in a normality-assuming estimation procedure? We still await some demonstration of which of these is the more serious error.

I recommend that researchers report both the original and new category boundaries or threshold values used whenever a polyserial or polychoric correlation matrix is used as input into LISREL. This would permit one's audience to assess the degree of adjusting that has been done, and provide a benchmark for the degree of compensatory readjusting that must be done if one is to interpret the LISREL model as accounting for the observed data, as opposed to the massaged data.[49] The seemingly unnoticed absence of guidelines for doing any compensatory readjustments when reporting effect estimates based on adjusted matrices suggests that more than a few people do not see any difference between accounting for real as opposed to adjusted data.

Those who have not found their current research concern enumerated above may be able to find assistance in the page of topicalized references provided by Bentler (1992a:12) or the annotated bibliography of nearly 300 items provided by Austin and Wolfle (1991).

No housekeeping is ever truly finished. I know there is more I could do, but I am exhausted. So I will rest, enjoy the semblance of order, and await the smudges from data-soiled hands, the crumbs of falling theories, the dust-balls of empiricism, all of which I will gladly endure in exchange for the happiness, joy, and laughter of those who enter herein.

> "While current methodological work is brilliant and clever, it is also ... arrogant in that anyone who dares to think a methodological nonconventional thought is told, in Greek letters and maximum likelihood estimators, to get lost." (Richard Zeller, 1987:409)

versus

> "May Greek letters and maximum likelihood estimators guide you to some nonconventional, clever, even brilliant, methodological thoughts. Respect will keep the resulting academic burden on your shoulders, so let friendly cheerfulness lighten the load." (Leslie Hayduk)

Notes

1. Though the discussion of the coefficient of determination appears under a major heading called "Assessment of Fit" (Joreskog and Sorbom, 1988:40) the subheading under which this appears implicitly claims that this is not a measure of fit per se, but is merely another coefficient that should be within a reasonable range before a solution should be accepted as reasonable. It is clear that the coefficient of determination does not provide an assessment of the fit of a model. There is nothing in the formula for the coefficient of determination that refers to the data, or the match of the model's implications to that data. Hence, the coefficient of determination is not an indicator of model fit and must be defended on other grounds.

2. One obtains the estimated numerical value of ψ_{33} from the ψ matrix, and the value of Var(η_3) from the output labelled "Covariances among the concepts." LISREL calculates this using Joreskog and Sorbom (1988: Eq. 1.26) or *Essentials* (Eq. 4.49).

3. See *Essentials* (70-73).

4. *Essentials* (71).

5. This increase will be bounded by a maximum of 1.0 given that the η's are standardized, unless the model contains a negative loop which might permit the error variance to exceed the variance of the associated concept as discussed in Chapter 3.

6. An error covariance contributes to the covariance between the two endogenous concepts, so we are imagining here that the preference is for an error covariance over any other way of providing a comparable degree of covariance between the concepts.

7. For a model with three η's with independent error variables, the coefficient of determination is

$$\text{CoD} = 1.0 - \frac{\begin{vmatrix} \psi_{11} & 0 & 0 \\ 0 & \psi_{22} & 0 \\ 0 & 0 & \psi_{33} \end{vmatrix}}{\begin{vmatrix} \text{Var}(\eta_1) & \text{Cov}(\eta_1\eta_2) & \text{Cov}(\eta_1\eta_3) \\ \text{Cov}(\eta_1\eta_2) & \text{Var}(\eta_2) & \text{Cov}(\eta_2\eta_3) \\ \text{Cov}(\eta_1\eta_3) & \text{Cov}(\eta_2\eta_3) & \text{Var}(\eta_3) \end{vmatrix}}.$$

The determinant of a diagonal matrix is the product of the diagonal elements, so the determinant of the numerator of the fraction is $\psi_{11}\psi_{22}\psi_{33}$. The determinant of the matrix in the denominator is more complex, and can be obtained by expanding the first row of the matrix (*Essentials*, 72). This provides

$$\begin{aligned} \text{Determinant of Denominator} = {} & \text{Var}(\eta_1) \begin{vmatrix} \text{Var}(\eta_2) & \text{Cov}(\eta_2\eta_3) \\ \text{Cov}(\eta_2\eta_3) & \text{Var}(\eta_3) \end{vmatrix} \\ & - \text{Cov}(\eta_1\eta_2) \begin{vmatrix} \text{Cov}(\eta_1\eta_2) & \text{Cov}(\eta_2\eta_3) \\ \text{Cov}(\eta_1\eta_3) & \text{Var}(\eta_3) \end{vmatrix} \\ & + \text{Cov}(\eta_1\eta_3) \begin{vmatrix} \text{Cov}(\eta_1\eta_2) & \text{Var}(\eta_2) \\ \text{Cov}(\eta_1\eta_3) & \text{Cov}(\eta_2\eta_3) \end{vmatrix} \end{aligned}$$

in which we thrice use the fact that the determinant of a 2x2 matrix is the product of the main diagonal elements minus the product of the other diagonal's elements, to get

$$= Var(\eta_1)[Var(\eta_2)Var(\eta_3) - Cov(\eta_2\eta_3)Cov(\eta_2\eta_3)]$$
$$- Cov(\eta_1\eta_2)[Cov(\eta_1\eta_2)Var(\eta_3) - Cov(\eta_1\eta_3)Cov(\eta_2\eta_3)]$$
$$+ Cov(\eta_1\eta_3)[Cov(\eta_1\eta_2)Cov(\eta_2\eta_3) - Cov(\eta_1\eta_3)Var(\eta_2)].$$

Multiplying out, collecting terms, and assuming that all the variables are standardized so their variances become 1.0 produces a determinant for the denominator of

$$= 1x1x1 + 2Cov(\eta_1\eta_2)Cov(\eta_1\eta_3)Cov(\eta_2\eta_3)$$
$$- Cov(\eta_1\eta_2)^2 - Cov(\eta_1\eta_3)^2 - Cov(\eta_2\eta_3)^2$$

which allows us to write the coefficient of determination for three variables with independent errors as

$$CoD = 1 - \frac{\psi_{11}\psi_{22}\psi_{33}}{1 + 2Cov(\eta_1\eta_2)Cov(\eta_1\eta_3)Cov(\eta_2\eta_3) - Cov(\eta_1\eta_2)^2 - Cov(\eta_1\eta_3)^2 - Cov(\eta_2\eta_3)^2}.$$

In this form, it is clear that the coefficient of determination for three concepts increases as the error variances decrease, again with the rider that if any one error variance becomes zero the coefficient of determination becomes 1.0. And what are we to make of the dependence of the coefficient of determination on the complex denominator? Some of these terms are necessarily positive (the squared covariances) but the triple-product term may be positive or negative depending on the signs of the three covariances. It will usually be informative to examine the individual concepts' error variances, but what is gained by expressing these variances as a fraction of the complex function of the covariances in the denominator? In particular, why should we prefer models having zero or two negative covariances over models implying one or three negative covariances between the concepts (as implied by the triple-product term)? I see no justification for this.

8. The coefficient of determination is not reported in the SIMPLIS version of LISREL 8, though it is unclear whether it is reported in the LISREL version of LISREL 8 (Joreskog and Sorbom, 1993).

9. For examples, see Section 6.3.

10. Kaplan (1988, 1989a) and Lance, Cornwell and Mulaik (1988) use base models resembling the model in Figure 6.1 in numerous respects, but the remaining differences between these models are too large to be treated as trivial, and hence the generalizability of test results across even so similar a set of models is questionable.

11. The Θ_ϵ for each indicator with a 1.0 λ (each single indicator and the "best" of the multiple indicators) should be given fixed values as discussed in Chapter 1.

12. Some reasonable constraints in the motif model might be $\beta_{65} = \beta_{64}$, $\beta_{65} = \beta_{56}$, or $\lambda_{86} = \lambda_{96}$.

13. Definitions can be awkward because they lead to colinearities. For example, if two independent variables, each with variance 1.0, are summed to define a "total" variable, the covariance matrix for the sum and other two variables looks like

$$\begin{bmatrix} 2.0 & 1.0 & 1.0 \\ 1.0 & 1.0 & 0.0 \\ 1.0 & 0.0 & 1.0 \end{bmatrix}.$$

When input into LISREL 7.20, this matrix produces a message that the matrix is not positive definite (since row 1 equals the sum of rows 2 and 3), and LISREL proceeded after inserting a ridge constant of .001. In contrast, inserting the data matrix

$$\begin{bmatrix} 2.00001 & 1.0 & 1.0 \\ 1.0 & 1.0 & 0.0 \\ 1.0 & 0.0 & 1.0 \end{bmatrix}$$

resulted in no error message regarding positive definiteness and produced model estimates much closer to the true model values than did the model based on the ridge option.

The insertion of a small colinearity-breaking variance increment can be used with either S or Σ. When used with S, this is similar to what is done by invoking the ridge option, though it is more conservative. The ridge option increases all the variances in the covariance matrix, not just one (Joreskog and Sorbom, 1988:25), and though it increases them by a reasonably small amount (.001 times each variance, Joreskog and Sorbom, 1988:127), it does not necessarily use the smallest amount possible. The default ridge option inserted for the example above was about 100 times what was necessary. Furthermore, we might anticipate that subtracting .00001 from the "total" variance would be equally effective at breaking the colinearity created by the definition, and would permit bracketing the range of variation in the estimates provided by our arbitrary intervention. The ridge option does not consider negative values.

Beckie (1994) presents an example where the resolution of a substantial methodological question required use of a definition, and Joreskog and Sorbom (1988:123-129) present an econometric example with multiple defined relationships.

Another option is to switch to using estimates created by unweighted least squares, since this estimation procedure does not require a positive definite data matrix (Wothke, 1993:265). One cost of adopting this approach is that no overall model χ^2 test is available for ULS estimation.

We should also note that including or excluding definitionally related variables has consequences for the χ^2 test. Because no new coefficients need to be estimated upon insertion of the definitional relation, the new line of data covariances adds to the degrees of freedom for the model χ^2 test (*Essentials*, 162). And the new data line can alter the estimates of the model coefficients and consequently alter the value of χ^2 itself. Any ill fit in the covariances pertaining to the variables involved in the definition become "doubly important," and hence are minimized at the expense of the fit of other covariances, because any ill fit in the component parts of a definition implies an additional ill fit to the covariances with the defined sum or difference. Consequently, the effects leading to or from the definitionally connected variables will be granted additional weight because they now carry unalterable implications for more data points. This implies there is a slight change in precisely what the model χ^2 is testing.

In general, the estimates from models including known definitional relations are better than estimates from models omitting the definitions because we can be assured that the inclusive model's estimates do not contradict the known definitional relations. Researchers should not overlook obvious or known relations among their modeled variables. If the relations are modeled, there is less possibility that the known relations will be violated by the compromise coefficient estimates, and one may even be able to capitalize on the known equation and additional data constraints to identify otherwise underidentified models.

14. ψ_{77} is underidentified unless η_7 influences at least two other variables (MacCallum and Browne, 1993:539), hence the need for either a fixed (and theoretically defended) non-zero value, or a third equality constraint to some other variance, for example to ψ_{99} or ψ_{44}. I think we can safely leave it up to the Monte Carlo experts to decide on the magnitudes of the parameters that are likely to prove most informative for

their particular application. The model is depicted in all-η form to maximize the output diagnostics and to facilitate comparisons with output from programs using other notation styles.

15. These include: Balderjahn (1988), Bentler (1990a,b, 1992a,b), Bentler and Mooijaart (1989), Bollen (1989b, 1990a), Bollen and Liang (1988), Bozdogan (1987), Cudeck and Henly (1991), Goffin (1993), Hu, Bentler and Kano (1992), Kaplan (1988), LaDu and Tanaka (1989), Maiti and Mukherjee (1990, 1991), Marcoulides (1990), Marsh and Balla (1994), Marsh, Balla and McDonald (1988), McDonald and Marsh (1990), Mulaik et al. (1989), Rindskopf and Rose (1988), Saris, Den Ronden and Satorra (1987), Satorra (1989), Tanaka and Huba (1989), Wheaton (1987), Wood and Brown (1994), and Yung and Bentler (1994).

And there are some disagreements. For example, the paper by Kaplan (1990) was followed by comments by MacCallum (1990), Bentler (1990b), Steiger (1990), Bollen (1990c), Tanaka (1990), Hayduk (1990) and a rejoinder by Kaplan (1990).

16. Recall *Essentials* (167-169). But note that there is a case to be made for the view that overall test statistics *should* be sensitive to *N* (Tanaka, 1993:32).

17. Browne and Cudeck (1993:146) recommend a test of close model fit, as opposed to exact model fit. They feel that the hypothesis of an exact model fit is "invariably false" and hence that some degree of close fit is a more appropriate benchmark. This leads them to make an arbitrary choice of .05 on the metric of the test statistic as being close enough "based on a substantial amount of experience with estimates" of this kind (Browne and Cudeck, 1993:146-147). The interesting thing about this is that the probabilities arising from this close as opposed to exact fit hypothesis are close to the probabilities one would get by simply using an *N* of 200 with the usual χ^2 (*Essentials*, 168). That is, simply ignoring the discriminating power provided by cases beyond 200, which corresponds to what is accomplished by Hoelter's critical N and which Hoelter (1983) also justified by an appeal to considerable research experience, provides about the right amount of discriminating power for the kinds of research enterprises these researchers have encountered. The question this prompts is, do we really need a new measure if all it is doing is reminding us that large *N*'s can make us reject models which are not all that bad?

Cudeck and Browne (1992) provide an alternative way of conceptualizing model fit appropriate for models that are conceived a priori as fitting only approximately in the population.

18. See Bollen (1989a:266) or *Essentials* (137, 160).

19. This will seem like a heavy penalty to researchers accustomed to estimating regression-like models where all the coefficients in the Φ matrix are free, though of considerably less interest than the effects on the dependent variable.

20. The justification for this is apparent in Eq. 6.8.

21. This advice arises in the context of Kaplan's noting that the difference between two nested models' definition two *AIC*'s should be approximated by the corresponding modification index minus twice the difference in the number of degrees of freedom. This suggests that *AIC* differences function like conservative modification indices.

22. Tanaka (1993), for example, favors indices that reward parsimony and indices that apply to different estimation methods, but he finds that method specific indices (such as the *AIC*, which depends on maximum likelihood procedures) function better. He also favors 0-1 normed indices and indices like χ^2 which depend on sample size. Tanaka views the issues of whether null models or substantive models should serve as a baseline, and whether the index should correspond to some population characteristic, as more controversial.

23. Discussion of the degrees of freedom for χ^2 and the segments of the model contributing to, or detracting from, the degrees of freedom address the issue of parsimony. As discussed in Section 2.3.1 (point 6), the χ^2 probability should be over .75, rather than over .05, for reasonable sized *N*'s, say up to about 400 or 500.

Even the suggestion to routinely report the AGFI is somewhat suspect because the sampling distribution of this index is still not known (Joreskog and Sorbom, 1988:43, 1993:123). My personal experience with the AGFI has provided me a strong and relatively precise criterion for this measure. Namely the AGFI for a model must be at least .95 before there is even a slim chance that the model will display no substantial signs of ill fit, and the AGFI will probably be .96 or .97 before the residuals and all the other output become symptom free. In research areas of substantive concern to me, I make mental note of authors sufficiently unobservant to publish models with .90 AGFI's.

24. Specifically the residuals, modification indices, explained variance, the magnitudes of particular model coefficients, standard errors, and signs of convergence problems.

25. Readers inclined to pursue a holy grail fit measure will find the definitions, interpretations, and properties of various indices in Tanaka (1993), Marsh, Balla and McDonald (1988), McDonald and Marsh (1990), and Joreskog and Sorbom (1993).

Segments of the LISREL 8 manual (Joreskog and Sorbom, 1993:117, 120, 123-124, 129-131) are intended to summarize a relatively new approach to assessing model fit. One key component is a noncentral χ^2 parameter λ, which is estimated, and for which a 90 percent confidence interval is reported. If the model is adequate, the confidence interval should reach zero. But only readers willing to volunteer for an honorary degree in statistics by studying Browne and Cudeck (1993) will appreciate the implicit conceptualization of adequacy being employed. Unfortunately, Joreskog and Sorbom (1993) are of limited assistance because they are unclear about the philosophical changes implicit in the new procedures. Browne and Cudeck (1993:157) warn that the fit indices they propose "should not be used in a mechanical decision process for selecting models" yet this is all the LISREL 8 reader is prepared for if all he or she has is the definition of yet another index, its confidence interval, and a target of 0.0.

26. See, for example, MacCallum (1986), MacCallum, Roznowski and Necowitz (1992), Chou and Bentler (1990), Kaplan (1989a), and Suyapa, Silvia and MacCallum (1988).

27. Chou and Bentler (1993) discuss a multivariate expected parameter change (MEPC) statistic and its standardized version (SMEPC).

28. The designation of this currently fixed coefficient as a β is purely for didactic specificity. An obvious parallel holds for any fixed coefficient, be it a β, γ, λ, ϕ, ψ, or θ.

29. The fit has been optimized by adjusting all the model's free coefficients to minimize the discrepancy between the data S and Σ, as discussed in *Essentials* (132-138). The β coefficient considered in Figure 6.3 has a fixed value of zero in this model and hence its value has not been altered during the parameter estimation process.

30. Kaplan (1989b:290), following Saris, Satorra and Sorbom (1987) calculate the *EPC* as the corresponding modification index divided by -(*N*-1) times the LISREL's corresponding "First Order Derivative."

31. Sorbom (1989) provides the statistical basis for the improvements in the modification index made between LISREL 5 and LISREL 6. The improvements make the modification index a more accurate prediction of the χ^2 change that is likely to arise from the freeing of a previously fixed or constrained coefficient by moving the x closer to the dot at the actual minimum, but the overall meaning and utility of the modification indices remain as illustrated.

32. Recall Chapter 4 and *Essentials* (177).

33. A χ^2 with one degree of freedom is significant at the .05 level if it exceeds 1.96^2, so modification indices less than 4.0 are unhelpful. To avoid capitalizing on chance one should only free coefficients with modification indices considerably above this, though there are too many relevant considerations to specify a standard cutoff for how much above 4.0 is enough.

34. The unreasonableness may reflect impossibilities such as negative variances, or the contradiction of the theoretically anticipated sign of the coefficient.

35. See *Essentials* (180-182).

36. Compare Kaplan (1989b) with *Essentials* (180-182).

37. Part of diligent searching is to make oneself sensitive to the things that modification indices and *EPC* coefficients often do not, but can be made to, do. These coefficients are calculated for all the fixed coefficients in a model's specification, but there may be numerous possible and reasonable coefficients that are not present in the current model specification and hence which are not assigned modification indices or *EPC* values. For example, if one uses the usual designation of a LISREL model with x's as indicators of the ξ concepts, and y's as indicators of the η concepts, direct effects from the ξ's to the y's, and direct effects from the η's the x's are not part of the model's representation, even though these may be reasonable additional model coefficients. A covariance between the errors on an x and a y variable are also not part of a model's representation in versions of LISREL prior to LISREL 8. One can obtain modification indices and *EPC* coefficients for all these types of coefficients in all versions of LISREL by adopting an all-η model specification (*Essentials*, 209-212). Similarly, more modification indices and *EPC* coefficients are reported if one avoids the use of DIagonal, IDentity, or ZEro specifications on LISREL's MOdel command line.

38. In general, I do not recommend eliminating insignificant coefficients from a model. If your reading of the literature led you to include these coefficients, there probably are many other researchers who will continue to be similarly misled unless you specifically make a point of reporting your inability to locate these effects. One instance where I would counsel the elimination of insignificant coefficients is if the identification of a more complex model becomes possible through elimination of the insignificant coefficient (see the example at the end of Chapter 5). Even in this context, only a minimum number of insignificant coefficients should be eliminated, and this should be accompanied by a clear textual report of all the eliminated insignificant coefficients.

It is not uncommon to find insignificant estimates for effects that were entered merely to provide statistical control for standard variables like age or income. If you believe both that such a variable is ineffective, and that statistical control is required to quiet some opposition, there is a tactic that can save you degrees of freedom, minimize identification problems, and reduce the number of coefficients, all without sacrificing the demonstration of statistical control. The trick is to enter the control variable in the model but to provide it with only the effects you really believe it displays, which may be none. If all the modification indices for the possible effects of the variable are small, then you can be assured that any effect estimates would have been near zero had they been estimated. Hence, you can interpret this as if the variable was in the equation, and hence controlled for, even though no degrees of freedom are wasted in demonstrating control for an ineffective variable. Naturally, any substantial (greater than 4.0) modification indices should be addressed, both in the model and in your report of the model.

39. Sobel alludes to other problems as well, but these too are often a function of his perspective, which is largely philosophical and econometric, and hence dissolve if one adopts a perspective similar to that implicit in the preceding chapters. For example, Sobel (1990) rigidly applies the jargon suggesting that locating and interpreting causal effects demands holding other variables constant. The problem this creates is that "'holding the other variables constant' does more than eliminate irrelevant variations in other portions of the model. It restricts the mechanisms by which a unit change can influence the dependent variable to only those routings that do not proceed through intervening variables or causal loops" (*Essentials*, 249). Sobel's rigid use of "holding constant" reinforces his thinking about segmenting models and their interpretations into those parts that do and do not permit free variation and hence causal interpretation.

But another solution is equally clear. It is not to partition the model into only those segments that avoid variables that are held constant (as Sobel suggests), but to modify the standard interpretation to avoid the undesirable partitioning implications of the phrase "holding constant." "It should now be clear why we routinely employ the phrase '*with the other variables left untouched*' rather than 'holding the other variables constant': 'holding constant' implies restrictions over and above those inherent in the model itself." (*Essentials*, 249, emphasis added). By permitting all the other model variables to be free to vary in response to, but to vary *only* in response to, a hypothetical unit change in a causal variable, the awkward segmentation that Sobel struggles with simply vanishes.

I view other aspects of Sobel (1990) as artifacts created because we have not yet fully realized that *the three traditional conditions for establishing causality are either false or misleading*. The condition that the cause and effect must be correlated is discredited by the simple demonstration that if a variable has a negative effect on another variable, and these variables each receive a positive effect from a common cause, the observed correlation can be zero despite the existence of the negative effect.

The "time sequence" condition (which Sobel refers to as Hume's criterion) is not consistent with reciprocal effects in models (cf. Sobel, 1990:506-507, *Essentials*, 250-253). By interpreting a looped model as representing a dynamic system that stabilizes prior to data collection, one avoids any need to preserve a pictorial or graphical temporal sequencing between the modeled variables, though models with loops or intervening variables do seem to require a reference to time in the sense that a single variable cannot function as both a cause and effect simultaneously. That is, coordination is required so that a variable is not acting as a cause at the same instant it is acting as an effect by changing.

The third standard condition for establishing causality, namely that the researchers provide evidence against a spurious relationship, is misleading because this has been superseded by a requirement for demonstrating the inability of multiple common sources and/or intervening effects to displace the direct effect, as is implicit in a full and accurately specified causal model. One needs a demonstration that even multiple common causes and multiple intervening variables are insufficient to supplant the particular causal connection being examined. Both spurious and intervening model segments must be eliminated before a direct causal effect is supported, *and there can be no rule as to how many spurious or how many intervening causes need to be controlled before what is left is definitively interpretable as a direct cause.*

40. Stacked models with differing sets of indicators, as per Chapter 5, might be of some help here.

41. Even some seemingly nonfactor MTMM models can be represented as second-order factor models within LISREL (Wothke and Browne, 1990).

42. Marsh, for example, reports the disconcerting observation that there are instances in which "so-called method factors contain substantial amounts of trait variance" (1989:353). And, as we saw in Chapter 3, even 10 repeated measures of a concept with a single method need not satisfy a factor model.

43. There are styles of multilevel models for which LISREL is not appropriate (DiPrete and Grusky, 1990).

44. Because the variances of $\xi_{\#}$ and $\xi_{\#+1}$ have been constrained to be equal, the nonindependent $\xi_{\#}$ segment contributes 3 squared, or 9, times the variance of the independent part $\xi_{\#+1}$, as can be seen by considering the equation for η_n as a special case of the variance equation for the model in column two of the preface figure. Note also that the model could have been drawn without η_n by permitting direct effects from $\xi_{\#}$ and $\xi_{\#+1}$ to η_5, and that this error segmenting procedure is equally applicable to ψ, Θ_{ϵ}, and Θ_{δ} error variances. If the segmental independence is thought to arise from connections between η_5's error to some of the model's other endogenous concepts or their error variables, the $\xi_{\#}$ and $\xi_{\#+1}$ in Figure 6.4 can be replaced with $\eta_{\#}$ and $\eta_{\#+1}$, and the variety of modeling possibilities this implies.

45. Without at least one covariance to another model ξ, the error partitioning variables $\xi_{\#}$ and $\xi_{\#+1}$ function interchangeably in the model, so the diagnostics for these variables should be the same except for scale differences.

46. Ipsative data arise when the sum of the values on a set of variables must equal a specific constant for every case. This occurs when entities are ranked (so the sum of the values for the sum of the variables recording the rankings for each case equals the sum of all the ranks because each of the ranks must be allocated to one of the entities), or when expenditures are expressed as a proportion of total income (so the sum of the expenditures is 1.0), or when a set of variables records whether the respondent voted for each of a political party or not (so the sum of these variables for each case is 1.0, as each case only has one vote to distribute among this list of possible parties).

47. See Aish and Joreskog (1990), Brown (1989), Homer and O'Brien (1988), Israels (1987), Joreskog (1990, 1994), Joreskog and Aish (forthcoming), Joreskog and Sorbom (1993), Lee and Leung (1992), Lee, Poon and Bentler (1990, 1992), Muthen (1993), O'Brien and Homer (1987), Poon and Lee (1987, 1992), Potthast (1993), Powers (1993), SPSS Inc. (1993), Wothke (1993:268), and Xie (1989).

48. Frequently only the endpoints and neutral midpoint of Likert scales are labeled, so that the respondents are forced to select from uniformly spaced numerical response options. Statisticians who believe that a response in the next higher category has an entirely flexible amount more of the relevant characteristic (e.g., Joreskog, 1994:383) are blinding themselves to the near(er) uniformity of the response spacings provided by such research practices, and hence are likely to overestimate the gains to be accrued when such observations are "corrected" for their mere ordinality.

49. This is not quite the same issue as whether the underlying multivariate normal latent variables PRELIS uses to do its adjusting need to be the same underlying latent variables (with the same meanings) as the latent variables in LISREL. In *Essentials* (330) I suggested that these had to be the same but I no longer support this view. The latent variables used by PRELIS in its derivation of the polyserial and polychoric correlations replace the *y* variables in LISREL, not any LISREL concepts. The easiest way to see this divergence is to imagine one PRELIS-adjusted indicator as arising from multiple concepts in LISREL. Under such a circumstance, there is no chance whatever for PRELIS's single assumed underlying normal variable to correspond to LISREL's multiple latent concepts. Joreskog (1994:383) follows Muthen in calling these "latent response variables." This phrase captures the latent nature of these variables in PRELIS and the indicator status of these very same variables in LISREL. This also hints at the issue raised in the text to which this note is attached, namely that the adjusted data in LISREL is now one unknown-sized step away from the real data unless the degree of adjusting done by PRELIS is reported.

Bibliography

Aaronson, Lauren S., Maureen A. Frey, and Carol J. Boyd. 1988. "Structural equation models and nursing research: Part II." *Nursing Research* 37:315-318.

Acock, Alan C. 1989. "Measurement error in secondary data analysis." Pp. 201-230 in Krishnan Namboodiri and Ronald C. Corwin (eds.) *Research in the Sociology of Education and Socialization: A Research Annual*. Greenwich, Conn.: JAI Press.

Aish, Anne-Marie, and Karl G. Joreskog. 1990. "A panel model for political efficacy and responsiveness: An application of LISREL 7 with weighted least squares." *Quality and Quantity* 24:405-426.

Ajzen, Icek, and Martin Fishbein. 1980. *Understanding Attitudes and Predicting Social Behavior*. Englewood Cliffs: Prentice-Hall.

Akaike, Hirotugu. 1987. "Factor analysis and AIC." *Psychometrika* 52:317-332.

Allison, Paul D. 1987. "Estimation of linear models with incomplete data." Pp. 71-103 in C. Clogg (ed.) *Sociological Methodology 1987*. San Francisco: Jossey-Bass.

Allison, Paul D., and Robert M. Hauser. 1991. "Reducing bias in estimates of linear models by remeasurement of a random subsample." *Sociological Methods and Research* 19:466-492.

Alwin, Duane. 1988. "Structural equation models in research on human development and aging." Pp. 71-170 in K. W. Schaie, R. T. Campbell, W. Meredith, and S. C. Rawling (eds.) *Methodological Issues in Aging Research*. New York: Springer.

Alwin, Duane F. 1989. "Problems in the estimation and interpretation of the reliability of survey data." *Quality and Quantity* 23:277-331.

Alwin, Duane F. 1992. "Information transmission in the survey interview: Number of response categories and the reliability of attitude measurement." Pp. 83-118 in Peter V. Marsden (ed.) *Sociological Methodology 1992*. Washington: American Sociological Association.

Alwin, Duane F., and Jon A. Krosnick. 1991. "The reliability of survey attitude measurement: The influence of question and respondent attitudes." *Sociological Methods and Research* 20:139-181.

Anderson, James C., and David W. Gerbing. 1982. "Some methods for respecifying measurement models to obtain unidimensional construct measurement." *Journal of Marketing Research* 19:453-460.

Anderson, James C., and David W. Gerbing. 1988. "Structural equation modeling in practice: A review and recommended two-step approach." *Psychological Bulletin* 103:411-423.

Anderson, James C., and David W. Gerbing. 1992. "Assumptions and comparative strengths of the two-step approach: Comment on Fornell and Yi." *Sociological Methods and Research* 20:321-333.

Anderson, James C., and James A. Narus. 1984. "A model of the distributor's perspective of the distributor-manufacturer working relationship." *Journal of Marketing* 48:62-74.

Arbuckle, James. 1988. *Getting started with AMOS under MSDOS*. Manuscript. Department of Psychology, Temple University, Philadelphia, Pennsylvania.

Arminger, Gerhard, and Ronald J. Schoenberg. 1989. "Pseudo maximum likelihood estimation and a test for misspecification in mean and covariance structure models." *Psychometrika* 54:409-425.

Armstrong, Paula S., and Michael D. Schulman. 1990. "Financial strain and depression among farm operators: The role of perceived economic hardship and personal control." *Rural Sociology* 55:475-493.

Arnetz, Bengt B., Sten-Olof Brenner, Lennart Levi, Robert Hjelm, Inga-Lill Petterson, Jerzy Wasserman, Bjorn Petrini, Peter Eneroth, Anders Kallner, Richard Kvetnansky, and Milan Vigas. 1991. "Neuroendocrine and immunologic effects of unemployment and job security." *Psychotherapy and Psychosomatics* 55:76-80.

Austin, James T., and Lee M. Wolfle. 1991. "Annotated bibliography of structural equation modelling: Technical work." *British Journal of Mathematical and Statistical Psychology* 44:93-152.

Avakame, Edem Frank. 1993. "Explaining domestic violence." Doctoral dissertation. Department of Sociology, University of Alberta, Edmonton, Alberta.

Baer, Douglas E., Edward Grabb, and William A. Johnston. 1991. "Economic dissatisfaction, potential unionism, and attitudes toward unions in Canada." *Canadian Review of Sociology and Anthropology* 28:67-83.

Balderjahn, Ingo. 1988. "A note on Bollen's alternative fit measure." *Psychometrika* 53:283-285.

Bandalos, Deborah L. 1993. "Factors influencing cross-validation of confirmatory factor analysis models." *Multivariate Behavioral Research* 28:351-374.

Batista-Foguet, Joan Manuel, and Willem E. Saris. 1992. "A new measurement procedure for attitudinal research: Analysis of its psychometric and informational properties." *Quality and Quantity* 26:127-146.

Beckie, Theresa. 1994. "The quality of life after coronary artery bypass graft surgery." Doctoral dissertation. Department of Educational Psychology, University of Alberta, Edmonton, Alberta.

Beland, Francois, and Brigitte Maheux. 1989. "Construct validity and second-order factorial model: The second-order model." *Quality and Quantity* 23:143-159.

Benson, Jeri, and Deborah L. Bandalos. 1992. "Second-order confirmatory factor analysis of the Reactions to Tests scale with cross-validation." *Multivariate Behavioral Research* 27:459-487.

Benson, Jeri, and John A. Fleishman. 1994. "The robustness of maximum likelihood and distribution-free estimators to non-normality in confirmatory factor analysis." *Quality and Quantity* 28:117-136.

Bentler, P. M. 1985. *Theory and Implications of EQS: A Structural Equations Program*. Los Angeles: BMDP Statistical Software.

Bentler, P. M. 1988. "Causal modeling via structural equation systems." Pp. 317-335 in John Nesselroade and Raymond Cattell (eds.) *Handbook of Multivariate Experimental Psychology*. 2nd ed. New York: Plenum.

Bentler, P. M. 1990a. "Comparative fit indexes in structural models." *Psychological Bulletin* 107:238-246.

Bentler, P. M. 1990b. "Fit indexes, Lagrange multipliers, constraint changes and incomplete data in structural equation models." *Multivariate Behavioral Research* 25:163-172.

Bentler, Peter M. 1992a. *EQS Structural Equations Program Manual*. Los Angeles: BMDP Statistical Software.

Bentler, Peter M. 1992b. "On the fit of models to covariances and methodology to the Bulletin. *Psychological Bulletin* 112:400-404.

Bentler, P. M., and Chih-Ping Chou. 1987. "Practical issues in structural modeling." *Sociological Methods and Research* 16:78-117.

Bentler, P. M., and Chih-Ping Chou. 1990. "Model search with Tetrad II and EQS." *Sociological Methods and Research* 19:67-79.

Bentler, P. M., and Chih-Ping Chou. 1992. "Some new covariance structure model improvement statistics." *Sociological Methods and Research* 21:259-282.

Bentler, P. M., and Chih-Ping Chou. 1993. "Some new covariance structure model improvement statistics." Pp. 235-255 in K. Bollen and R. Stine (eds.) *Testing Structural Equation Models*. Newbury Park: Sage.

Bentler, P. M., and Theo Dijkstra. 1985. "Efficient estimation via linearization in structural models." Pp. 9-42 in P. R. Krishnaiah (ed.) *Multivariate Analysis VI*. New York: North-Holland.

Bentler, P. M., and Ab Mooijaart. 1989. "Choice of structural model via parsimony: A rationale based on precision." *Psychological Bulletin* 106:315-317.

Berk, Richard A. 1991. "Toward a methodology for mere mortals." Pp. 315-323 in Peter V. Marsden (ed.) *Sociological Methodology 1991*. Washington: American Sociological Association.

Berkane, Maia, and P. M. Bentler. 1988. "Estimation of contamination parameters and identification of outliers in multivariate data." *Sociological Methods and Research* 17:55-64.

Blalock, Hubert M. Jr. 1964. *Causal Inference in Nonexperimental Research*. Chapel Hill: University of North Carolina Press.

Blalock, Hubert M. Jr. 1979. *Social Statistics*. Rev. 2d. ed. New York: McGraw-Hill.

Blalock, Hubert M. Jr. 1991. "Are there really any *constructive* alternatives to causal modeling?" Pp. 325-335 in Peter V. Marsden (ed.) *Sociological Methodology 1991*. Washington: American Sociological Association.

Blanchard, Ray. 1994. "A structural equation model for age at clinical presentation in nonhomosexual male gender dysphorics." *Archives of Sexual Behavior* 23:311-320.

Blanchard-Fields, Freda, Lynda Suhrer-Roussel, and Christopher Hertzog. 1994. "A confirmatory factor analysis of the Bem Sex Role Inventory: Old questions, new answers." *Sex Roles* 30:423-458.

Bollen, Kenneth A. 1987a. "Outliers and improper solutions: A confirmatory factor analysis example." *Sociological Methods and Research* 15:375-384.

Bollen, Kenneth A. 1987b. "Total, direct, and indirect effects in structural equation models." Pp. 37-69 in Clifford C. Clogg (ed.) *Sociological Methodology 1987*. Washington: American Sociological Association.

Bollen, Kenneth A. 1989a. *Structural Equations with Latent Variables*. New York: Wiley.

Bollen, Kenneth A. 1989b. "A new incremental fit index for general structural equation models." *Sociological Methods and Research* 17:303-316.

Bollen, Kenneth A. 1990a. "Overall fit in covariance structure models: Two types of sample size effects." *Psychological Bulletin* 107:256-259.

Bollen, Kenneth A. 1990b. "Outlier screening and a distribution-free test for vanishing tetrads." *Sociological Methods and Research* 19:80-92.

Bollen, Kenneth A. 1990c. "A comment on model evaluation and modification." *Multivariate Behavioral Research* 25:181-185.

Bollen, Kenneth A., and Gerhard Arminger. 1991. "Observational residuals in factor analysis and structural equation modeling." Pp. 235-262 in Peter V. Marsden (ed.) *Sociological Methodology 1991*. Washington: American Sociological Association.

Bollen, Kenneth A., and Richard Lennox. 1991. "Conventional wisdom on measurement: A structural equation perspective." *Psychological Bulletin* 110:305-315.

Bollen, Kenneth A., and Jersey Liang. 1988. "Some properties of Hoelter's CN." *Sociological Methods and Research* 16:492-503.

Bollen, Kenneth A., and J. Scott Long. 1992. "Tests for structural equation models: Introduction. *Sociological Methods and Research* 21:123-131.

Bollen, Kenneth A., and J. Scott Long (eds.). 1993. *Testing Structural Equation Models*. Newbury Park: Sage.

Bollen, Kenneth A., and Robert Stine. 1990. "Direct and indirect effects: Classical and bootstrap estimates of variability." Pp. 115-140 in Clifford Clogg (ed.) *Sociological Methodology 1990*. Washington: American Sociological Association.

Bollen, Kenneth A., and Robert A. Stine. 1992. "Bootstrapping goodness-of-fit measures in structural equation models." *Sociological Methods and Research* 21:205-229.

Bollen, Kenneth A., and Robert A. Stine. 1993. "Bootstrapping goodness-of-fit measures in structural equation models." Pp. 111-135 in K. Bollen and R. Stine (eds.) *Testing Structural Equation Models*. Newbury Park: Sage.

Bollen, Kenneth A., and Kwok-fai Ting. 1993. "Confirmatory TETRAD analysis." Pp. 147-175 in Peter V. Marsden (ed.) *Sociological Methodology 1993*. Washington: American Sociological Association.

Boomsma, Anne. 1987. "The robustness of maximum likelihood estimation in structural equation models." Pp. 160-188 in Peter Cuttance and Russell Ecob (eds.) *Structural Modeling by Example*. New York: Cambridge University Press.

Botvin, Gilbert J., Eli Baker, Elizabeth M. Botvin, Linda Dusenbury, John Cardwell, and Tracy Diaz. 1993. "Factors promoting cigarette smoking among black youth: A causal modeling approach." *Addictive Behaviors* 18:397-405.

Boyd, Carol J., Maureen A. Frey, and Lauren S. Aaronson. 1988. "Structural equation models and nursing research: Part I." *Nursing Research* 37:249-252.

Bozdogan, Hamparsum. 1987. "Model selection and Akaike's information criteria (AIC): The general theory and its analytical extensions." *Psychometrika* 52:345-370.

Breckler, Steven J. 1984. "Empirical validation of affect, behavior, and cognition as distinct components of attitude." *Journal of Personality and Social Psychology* 47:1191-1205.

Breckler, Steven J. 1990. "Applications of covariance structure modeling in psychology: Cause for concern." *Psychological Bulletin* 107:260-273.

Brown, R. L. 1989. "Using covariance modeling for estimating reliability on scales with ordered polytomous variables." *Educational and Psychological Measurement* 49:385-398.

Browne, Michael W. 1992. "Circumplex models for correlation matrices." *Psychometrika* 57:469-497.

Browne, M. W., and R. Cudeck. 1989. "Single sample cross-validation indices for covariance structures." *Multivariate Behavioral Research* 24:445-455.

Browne, Michael W., and Robert Cudeck. 1992. "Alternative ways of assessing model fit." *Sociological Methods and Research* 21:230-258.

Browne, Michael W., and Robert Cudeck. 1993. "Alternative ways of assessing model fit." Pp. 136-162 in K. Bollen and R. Stine (eds.) *Testing Structural Equation Models*. Newbury Park: Sage.

Browne, M. W., and S. H. C. DuToit. 1992. "Automated fitting of nonstandard models." *Multivariate Behavioral Research* 27:269-300.

Browne, M. W., and A. Shapiro. 1988. "Robustness of normal theory methods in the analysis of linear latent variate models." *British Journal of Mathematical and Statistical Psychology* 41:193-208.

Buhrich, Neil J., Michael Bailey, and Nicholas G. Martin. 1991. "Sexual orientation, sexual identity, and sex-dimorphic behaviors in male twins." *Behavior Genetics* 21:75-96.

Burt, Ronald S. 1973. "Confirmatory factor-analytic structures and the theory construction process (plus corrigenda)." *Sociological Methods and Research* 2:131-190.

Burt, Ronald S. 1976. "Interpretational confounding of unobserved variables in structural equation models." *Sociological Methods and Research* 5:3-51.

Byrne, Barbara M. 1988. "Measuring adolescent self-concept: Factorial validity and equivalency of the SDQ-III across gender." *Multivariate Behavioral Research* 23:361-375.

Byrne, Barbara M. 1989a. *A Primer of LISREL: Basic Applications and Programming for Confirmatory Factor Analytic Models*. New York: Springer-Verlag.

Byrne, Barbara M. 1989b. "Multigroup comparisons and the assumption of equivalent construct validity across groups: Methodological and substantive issues." *Multivariate Behavioral Research* 24:503-523.

Byrne, Barbara M. 1994. *Structural Equation Modeling with EQS and EQS/Windows: Basic Concepts, Applications, and Programming*. Thousand Oaks, CA: Sage.

Byrne, Barbara M., and Richard D. Goffin. 1993. "Modeling MTMM data from additive and multiplicative covariance structures: An audit of construct validity concordance. *Multivariate Behavioral Research* 28:67-96.

Byrne, Barbara M., Richard J. Shavelson, and Bengt Muthen. 1989. "Testing for the equivalence of factor covariance and mean structures: The issue of partial measurement invariance." *Psychological Bulletin* 105:456-466.

Camstra, Astrea, and Anne Boomsma. 1992. "Cross-validation in regression and covariance structure analysis." *Sociological Methods and Research* 21:89-115.

Chan, Jason C. 1991. "Response-order effects in Likert-type scales." *Educational and Psychological Measurement* 51:531-540.

Chan, Wai, and Peter M. Bentler. 1993. "The covariance structure analysis of ipsative data." *Sociological Methods and Research* 22:214-247.

Chipuer, Heather M., Michael J. Rovine, and Robert Plomin. 1990. "LISREL modeling: Genetic and environmental influences on IQ revisited." *Intelligence* 14:11-29.

Chou, Chih-Ping, and P. M. Bentler. 1990. "Model modification in covariance structure modeling: A comparison among likelihood ratio, Lagrange multiplier and Wald tests." *Multivariate Behavioral Research* 25:115-136.

Chou, Chih-Ping, and P. M. Bentler. 1993. "Invariant standardized estimated parameter change for model modification in covariance structure analysis." *Multivariate Behavioral Research* 28:97-110.

Chowdhury, Fakhrul Islam. 1991. "Fertility in Bangladesh: explanations through structural equation models." Doctoral dissertation. Department of Sociology, University of Alberta, Edmonton, Alberta.

Church, A. Timothy, and Peter J. Burke. 1994. "Exploratory and confirmatory tests of the Big Five and Tellegen's Three- and Four-Dimensional models." *Journal of Personality and Social Psychology* 66:93-114.

Clarkson-Smith, Louise, and Alan A. Hartley. 1990. "Structural equation models of relationships between exercise and cognitive abilities." *Psychology and Aging* 5:437-446.

Cliff, Norman. 1983. "Some cautions concerning the application of causal modeling methods." *Multivariate Behavioral Research* 18:115-126.

Cohen, Patricia, Jacob Cohen, Jeanne Teresi, Margaret Marchi, and Noemi Velez. 1990. "Problems in the measurement of latent variables in structural equations causal models." *Applied Psychological Measurement* 14:183-196.

Cole, David A. 1987. "Utility of confirmatory factor analysis in test validation research." *Journal of Consulting and Clinical Psychology* 55:584-594.

Cudeck, Robert. 1989. "Analysis of correlation matrices using covariance structure models." *Psychological Bulletin* 105:317-327.

Cudeck, Robert, and Michael W. Browne. 1992. "Constructing a covariance matrix that yields a specified minimizer and a specified minimum discrepancy function value." *Psychometrika* 57:357-369.

Cudeck, Robert, and S. J. Henly. 1991. "Model selection in covariance structures analysis and the 'problem' of sample size: A clarification." *Psychological Bulletin* 109:512-519.

Cudeck, Robert, Kelli J. Klebe, and Susan J. Henly. 1993. "A simple Gauss-Newton procedure for covariance structure analysis with high-level computer languages." *Psychometrika* 58:211-232.

Cui, Geng, and Sjef van den Berg. 1991. "Testing the construct validity of intercultural effectiveness." *International Journal of Intercultural Relations* 15:227-241.

Cuttance, Peter, and Russell Ecob. 1987. *Structural Modeling by Example: Applications in Educational, Sociological, and Behavioral Research.* New York: Cambridge University Press.

De Graaf, Nan Dirk, Jacques Hagenaars, and Ruud Luijkx. 1989. "Intragenerational stability of postmaterialism in Germany, the Netherlands and the United States." *European Sociological Review* 5:183-201.

DeMaio-Esteves, Maureen. 1990. "Mediators of daily stress and perceived health status in adolescent girls." *Nursing Research* 39:360-363.

Dijkstra, T. K. 1990. "Some properties of estimated scale invariant covariance structures." *Psychometrika* 55:327-336.

DiPrete, Thomas A., and Jerry D. Forristal. 1994. "Multilevel models: Methods and substance." *Annual Review of Sociology* 20:331-357.

DiPrete, Thomas A., and David B. Grusky. 1990. "The multilevel analysis of trends with repeated cross-sectional data." Pp. 337-368 in Clifford Clogg (ed.) *Sociological Methodology 1990*. Washington: American Sociological Association.

Dolan, Conor V., and Peter C. M. Molenaar. 1991. "A comparison of four methods of calculating standard errors of maximum-likelihood estimates in the analysis of covariance structure." *British Journal of Mathematical and Statistical Psychology* 44:359-368.

Dolan, Conor V., and Peter C. M. Molenaar. 1994. "Testing specific hypotheses concerning latent group differences in multi-group covariance structure analysis with structured means." *Multivariate Behavioral Research* 29:203-222.

Dolan, Conor V., Peter C. M. Molenaar, and Dorret I. Boomsma. 1992. "Decomposition of multivariate phenotypic means in multigroup genetic covariance structure analyses." *Behavior Genetics* 22:319-335.

Duncan, Otis Dudley. 1975. *Introduction to Structural Equation Models*. New York: Academic Press.

Dunlap, William P., and John M Cornwell. 1994. "Factor analysis of ipsative measures." *Multivariate Behavioral Research* 29:115-126.

Ellis, Jules L. 1993. "Subpopulation invariance of patterns in covariance matrices." *British Journal of Mathematical and Statistical Psychology* 46:231-254.

Embree, Bryan G., and Paul C. Whitehead. 1993. "Validity and reliability of self-reported drinking behavior: Dealing with the problem of response bias." *Journal of Studies on Alcohol* 28:334-344.

Entwisle, Doris R., Leslie A. Hayduk and Thomas W. Reilly. 1982. *Early Schooling: Cognitive and Affective Outcomes.* Baltimore: Johns Hopkins University Press.

Essentials. See Hayduk, L. 1987.

Farley, John U., and Srinivas K. Reddy. 1987. "A factorial evaluation of effects of model specification and error on parameter estimation in a structural equation model." *Multivariate Behavioral Research* 22:71-90.

Fassinger, Ruth E. 1987. "Use of structural equation modeling in counseling psychology research." *Journal of Counseling Psychology* 34:425-436.

Fleishman, John, and Jeri Benson. 1987. "Using LISREL to evaluate measurement models and scale reliability." *Educational and Psychological Measurement* 47:925-939.

Fornell, Claes, and Roland T. Rust. 1989. "Incorporating prior theory in covariance structure analysis: A Bayesian approach." *Psychometrika* 54:249-259.

Fornell, Claes, and Youjae Yi. 1992a. "Assumptions of the two-step approach to latent variable modeling." *Sociological Methods and Research* 20:291-320.

Fornell, Claes, and Youjae Yi. 1992b. "Assumptions of the two-step approach: Reply to Anderson and Gerbing." *Sociological Methods and Research* 20:334-339.

Fraser, Colin. 1988. *COSAN User's Guide.* Centre for Behavioural Studies in Education, University of New England, Armidale, Australia.

Fraser, Colin, and Roderick P. McDonald. 1988. "COSAN: Covariance structure analysis." *Multivariate Behavioral Research* 23:263-265.

Freedman, D. A. 1987. "As others see us: A case study in path analysis." *Journal of Educational Statistics* 12:101-128, with a rejoinder after comments 206-223.

Freedman, David A. 1991. "Statistical models and shoe leather." Pp. 291-313 in Peter Marsden (ed.) Sociological Methodology 1991. Washington: American Sociological Association.

Gallini, Joan K., and Jim F. Casteel. 1987. "An inquiry into the effects of outliers on estimates of a structural equation model of basic skills assessment." Pp. 189-201 in Peter Cuttance and Russell Ecob (eds.) *Structural Modeling by Example*. New York: Cambridge University Press.

Gerbing, David W., and James C. Anderson. 1985. "The effects of sampling error and model characteristics on parameter estimation for maximum likelihood confirmatory factor analysis." *Multivariate Behavioral Research* 20:255-271.

Gerbing, David W., and James C. Anderson. 1987. "Improper solutions in the analysis of covariance structures: Their interpretability and a comparison of alternate respecifications." *Psychometrika* 52:99-111.

Gerbing, David W., and James C. Anderson. 1992. "Monte Carlo evaluations of goodness-of-fit indices for structural equation models." *Sociological Methods and Research* 21:132-160.

Gerbing, David W., and James C. Anderson. 1993. "Monte Carlo evaluations of goodness-of-fit indices for structural equation models." Pp. 40-65 in K. Bollen and R. Stine (eds.) *Testing Structural Equation Models*. Newbury Park: Sage.

Germain, Guy. 1994. "Forced to choose and forced to care: Temporal tensions of 'women in the middle.'" Master's thesis. Department of Sociology, University of Alberta, Edmonton, Alberta.

Gillespie, Michael W. 1991. "Conceptual issues in models of sibling resemblance: Comment on Hauser and Mossel (1985) and Hauser (1988). *American Journal of Sociology* 97:196-206.

Gillespie, Michael W., and Janet L. McDonald. 1991. "Structural equation models of multiple respondent data: Mutual influence versus family factor approaches." Research Discussion Paper Number 83. Population Research Laboratory, Department of Sociology, University of Alberta, Edmonton, Alberta.

Girodo, Michel. 1991. "Personality, job stress, and mental health in undercover agents: A structural equation analysis." *Journal of Social Behavior and Personality* 6:375-390.

Glymour, Clark, Richard Scheines, and Peter Spirtes. 1988. "Exploring causal structure with the TETRAD program." Pp. 411-448 in Clifford C. Clogg (ed.) *Sociological Methodology 1988*. Washington: American Sociological Association.

Glymour, Clark, Richard Scheines, Peter Spirtes, and Kevin Kelly. 1987. *Discovering Causal Structure: Artificial Intelligence, Philosophy of Science and Statistical Modeling*. New York: Academic Press.

Glymour, Clark, Richard Scheines, Peter Spirtes, and Kevin Kelly. 1988. "TETRAD: Discovering causal structure." *Multivariate Behavioral Research* 23:279-280.

Glymour, Clark, and Peter Spirtes. 1988. "Latent variables, causal models and overidentifying constraints." *Journal of Econometrics* 39:175-198.

Glymour, Clark, Peter Spirtes, and Richard Scheines. 1990. "Independence relations produced by parameter values in causal models." *Philosophical Topics* 18:55-70.

Glymour, C., P. Spirtes, and R. Scheines. 1991a. "Causal inference." *Erkenntnis* 35:151-189.

Glymour, Clark, Peter Spirtes, and Richard Scheines. 1991b. "Inferring causal structure in mixed populations." Manuscript. Department of Philosophy, Carnegie-Mellon University, Pittsburgh, Pennsylvania.

Godwin, Deborah D. 1988. "Causal modeling in family research." *Journal of Marriage and the Family* 50:917-927.

Goffin, Richard D. 1993. "A comparison of two new indices for the assessment of fit of structural equation models." *Multivariate Behavioral Research* 28:205-214.

Goldstein, Harvey, and Roderick P. McDonald. 1988. "A general model for the analysis of multilevel data." *Psychometrika* 53:455-468.

Gollob, Harry F., and Charles S. Reichardt. 1985. "Building time lags into causal models of cross-sectional data." Pp. 165-170 in *Proceedings of the Social Statistics Section of the American Statistical Association*. Washington: American Statistical Association.

Gollob, Harry F., and Charles S. Reichardt. 1987. "Taking account of time lags in causal models." *Child Development* 58:80-92.

Golob, T. F., and H. Meurs. 1988. "Modeling the dynamics of passenger travel demand by using structural equations." *Environment and Planning A* 20:1197-1218.

Grayson, David, and Herbert W. Marsh. 1994. "Identification with deficient rank loading matrices in confirmatory factor analysis: Multitrait-multimethod models." *Psychometrika* 59:121-134.

Green, Donald Philip, and Bradley L. Palmquist. 1991. "More 'tricks of the trade': Reparameterizing LISREL models using negative variances." *Psychometrika* 56:137-145.

Greenberg, David F., and Ronald C. Kessler. 1982. "Equilibrium and identification in linear panel models." *Sociological Methods and Research* 10:435-451.

Gulick, Elsie E. 1992. "Model for predicting work performance among persons with multiple sclerosis." *Nursing Research* 41:266-272.

Gustafsson, Jan-Eric, and Gudrun Balke. 1993. "General and specific abilities as predictors of school achievement." *Multivariate Behavioral Research* 28:407-434.

Hagan, John, Celesta Albonetti, Duane Alwin, A. R. Gillis, John Hewitt, Alberto Palloni, Patricia Parker, Ruth Peterson, and John Simpson. 1989. *Structural Criminology*. New Brunswick, NJ: Rutgers University Press.

Hagan, John, and Blair Wheaton. 1993. "The search for adolescent role exits and the transition to adulthood." *Social Forces* 71:955-980.

Hakkinen, Unto. 1991. "The production of health and the demand for health care in Finland." *Social Science and Medicine* 33:225-237.

Harnqvist, Kjell, Jan-Eric Gustafsson, Bengt O. Muthen, and Ginger Nelson. 1994. "Hierarchical models of ability at individual and class levels." *Intelligence* 18:165-187.

Hartmann, W. M. 1992. *The CALIS procedure: Extended user's guide*. Cary, NC: SAS Institute.

Hattie, John. 1985. "Methodological review: Assessing unidimensionality of tests and items." *Applied Psychological Measurement* 9:139-164.

Hayduk, Leslie A. 1976. "Dimensions of personal space." Doctoral dissertation. Department of Social Relations, Johns Hopkins University, Baltimore.

Hayduk, Leslie A. 1985. "Personal space: The conceptual and measurement implications of structural equation models." *Canadian Journal of Behavioral Science* 17:140-149.

Hayduk, Leslie A. 1987. *Structural Equation Modeling with LISREL: Essentials and Advances*. Baltimore: Johns Hopkins University Press.

Hayduk, Leslie A. 1990. "Should model modifiactions be oriented toward improving data fit or encouraging creative and analytical thinking?" *Multivariate Behavioral Research* 25:193-196.

Hayduk, Leslie A. 1994. "Personal space: Understanding the simplex model." *Journal of Nonverbal Behavior* 18:245-260.

Hayduk, Leslie A., and Edem Frank Avakame. 1990/91. "Modeling the deterrent effect of sanctions on family violence: Some variations on the deterrence theme." *Criminometrica* 6/7:19-37.

Hayduk, Leslie A., Pamela A. Ratner, Joy L. Johnson, and Joan L. Bottorff. 1995. "Attitudes, ideology, and the factor model." *Journal of Political Psychology* 16:479-507.

Hays, Ron D., and Ken White. 1987. "The importance of considering alternative structural equation models in evaluation research." *Evaluation and the Health Professions* 10:90-100.

Henly, Susan J. 1993. "Robustness of some estimators for the analysis of covariance structures." *British Journal of Mathematical and Statistical Psychology* 46:313-338.

Herting, Jerald R., and Herbert L. Costner. 1985. "Respecification in multiple indicator models." Pp. 321-393 in H. M. Blalock (ed.) *Causal Models in the Social Sciences*. 2nd ed. New York: Aldine.

Hewitt, J. K., J. L. Silberg, M. C. Neale, L. J. Eaves, and M. Erickson. 1992. "The analysis of parental ratings of children's behavior using LISREL." *Behavior Genetics* 22:293-317.

Hines, Melissa, Lee Chiu, Lou Ann McAdams, Peter M. Bentler, and Jim Lipcamon. 1992. "Cognition and the corpus callosum: Verbal fluency, visiospatial ability, and language lateralization related to midsagittal surface areas of callocal subregions." *Behavioral Neuroscience* 106:3-14.

Hoelter, Jon W. 1983. "The analysis of covariance structures: Goodness-of-fit indices." *Sociological Methods and Research* 11:325-344.

Holahan, Charles J., and Rudolf H. Moos. 1991. "Life stressors, personal and social resources, and depression: A 4-year structural model." *Journal of Abnormal Psychology* 100:31-38.

Homer, Pamela, and Robert M. O'Brien. 1988. "Using LISREL models with crude rank category measures." *Quality and Quantity* 22:191-201.

Horton, Raymond L., and Patricia J. Horton. 1991. "A model of willingness to become a potential organ donor." *Social Science and Medicine* 33:1037-1051.

Hoyle, Rick H. forthcoming. *Structural Equation Modeling: Concepts, Issues, and Applications*. Newbury Park: Sage.

Hu, Li-tze, P. M. Bentler, and Yutaka Kano. 1992. "Can test statistics in covariance structure analysis be trusted?" *Psychological Bulletin* 112:351-362.

Hunter, John E., and David W. Gerbing. 1982. "Unidimensional measurement, second order factor analysis, and causal models." *Research in Organizational Behavior* 4:267-320.

Israels, A. Z. 1987. "Path analysis for mixed qualitative and quantitative variables." *Quality and Quantity* 21:91-102.

Iverson, Roderick D., and Parimal Roy. 1994. "A causal model of behavioral commitment: Evidence from a study of Australian blue-collar employees." *Journal of Management* 20:15-41.

Jaccard, James, and Choi K. Wan. 1995. "Measurement error in the analysis of interaction effects between continuous predictors using multiple regression: Multiple indicator and structural equation approaches." *Psychological Bulletin* 117:348-357.

Jagodzinski, Wolfgang, Steffen M. Kuhnel, and Peter Schmidt. 1988. "Is the true score model or the factor model more appropriate? Response to Saris and Putte." *Sociological Methods and Research* 17:158-164.

Jagodzinski, Wolfgang, Steffen M. Kuhnel, and Peter Schmidt. 1990. "Searching for parsimony: Are true-score models or factor models more appropriate?" *Quality and Quantity* 24:447-470.

James, Lawrence R., Stanley A. Mulaik, and Jeanne M. Brett. 1982. *Causal Analysis: Assumptions, Models and Data*. Beverly Hills: Sage.

Jenkins, J. Craig, and Augustine J. Kposowa. 1990. "Explaining military coups d'etat: Black Africa, 1957-1984." *American Sociological Review* 55:861-875.

Jensen, Arthur R., and Li-Jen Weng. 1994. "What is a good g?" *Intelligence* 18:231-258.

Johnson, Joy L., Pamela A. Ratner, Joan L. Bottorff, and Leslie A. Hayduk. 1993. "An exploration of Pender's health promotion model using LISREL." *Nursing Research* 42:132-138.

Joreskog, Karl G. 1990. "New developments in LISREL: Analysis of ordinal variables using polychoric correlations and weighted least squares." *Quality and Quantity* 24:387-404.

Joreskog, Karl G. 1993. "Testing structural equation models." Pp. 294-316 in K. Bollen and R. Stine (eds.) *Testing Structural Equation Models*. Newbury Park: Sage.

Joreskog, Karl G. 1994. "On the estimation of polychoric correlations and their asymptotic covariance matrix." *Psychometrika* 59:381-389.

Joreskog, Karl G., and Aish, A. M. forthcoming. *Structural Equation Modeling with Ordinal Variables*. Book manuscript in preparation.

Joreskog, Karl G., and Dag Sorbom. 1980. "Simultaneous analysis of longitudinal data from several cohorts." Research Report 80-5. Department of Statistics, University of Uppsala, Uppsala, Sweden.

Joreskog, Karl G., and Dag Sorbom. 1981. *LISREL V: Analysis of Linear Structural Relationships by Maximum Likelihood and Least Squares Methods*. Chicago: International Educational Services.

Joreskog, Karl G., and Dag Sorbom. 1988. *LISREL 7: A Guide to the Program and Applications*. Chicago: SPSS Inc.

Joreskog, Karl G., and Dag Sorbom. 1989. *LISREL 7: A Guide to the Program and Applications*. 2nd ed. Chicago: SPSS Inc.

Joreskog, Karl G., and Dag Sorbom. 1990. "Model search with Tetrad II and LISREL." *Sociological Methods and Research* 19:93-106.

Joreskog, Karl G., and Dag Sorbom. 1993. *LISREL 8: Structural Equation Modeling with the SIMPLIS Command Language*. Chicago: Scientific Software International.

Kandel, Eric R., James H. Schwartz, and Thomas M. Jessell. 1991. *Principles of Neural Science*. 3rd ed. New York: Elsevier.

Kaplan, David. 1988. "The impact of specification error on the estimation, testing, and improvement of structural equation models." *Multivariate Behavioral Research* 23:69-86.

Kaplan, David. 1989a. "A study of the sampling variability and Z-values of parameter estimates from misspecified structural equation models." *Multivariate Behavioral Research* 24:41-57.

Kaplan, David. 1989b. "Model modification in covariance structure analysis: Application of the expected parameter change statistic." *Multivariate Behavioral Research* 24:285-305.

Kaplan, David. 1989c. "Power of the likelihood ratio test in multiple group confirmatory factor analysis under partial measurement invariance." *Educational and Psychological Measurement* 49:579-586.

Kaplan, David. 1990. "Evaluating and modifying covariance structure models: A review and recommendation." *Multivariate Behavioral Research* 25:137-155, with rejoinder after comments 197-204.

Kaplan, David. 1991. "On the modification and predictive validity of covariance structure models." *Quality and Quantity* 25:307-314.

Kaplan, David. 1994. "Estimator conditioning diagnostics for covariance structure models." *Sociological Methods and Research* 23:200-229.

Kaplan, David, and R. Neil Wenger. 1993. "Asymptotic independence and separability in covariance structure models: Implications for specification error, power, and model modification." *Multivariate Behavioral Research* 28:467-482.

Kaplan, Howard B., and Robert J. Johnson. 1991. "Negative social sanctions and juvenile delinquency: Effects of labeling in a model of deviant behavior." *Social Science Quarterly* 72:98-122.

Keith, T. Z. 1990. "Confirmatory and hierarchical confirmatory analysis of the differential ability scales." *Journal of Psychoeducational Assessment* 8:391-405.

Kelderman, Henk. 1987. "LISREL models for inequality constraints in factor and regression analysis." Pp. 221-240 in Peter Cuttance and Russell Ecob (eds.) *Structural Modeling by Example*. New York: Cambridge University Press.

Kenny, David A., and Deborah A. Kashy. 1992. "Analysis of the multitrait-multimethod matrix by confirmatory factor analysis." *Psychological Bulletin* 112:165-172.

Kinzel, Cliff. 1993. "The Alberta Survey 1993 Sampling Report." Alberta/Edmonton Series Report No. 78. Population Research Laboratory, Department of Sociology, University of Alberta, Edmonton Alberta.

Kleck, Gary. 1991. *Point Blank: Guns and Violence in America*. New York: Aldine.

Koslowsky, Meni, Avi Bleich, Alan Apter, Zahava Solomon, Benny Wagner, and Alexander Greenspoon. 1992. "Structural equation modelling of some of the determinants of suicide risk." *British Journal of Medical Psychology* 65:157-165.

Kuhnel, Steffen M. 1988. "Testing MANOVA designs with LISREL." *Sociological Methods and Research* 16:504-523.

Kumar, Ajith, and William R. Dillon. 1987. "The interaction of measurement and structure in simultaneous equation models with unobservable variables." *Journal of Marketing Research* 24:98-105.

LaDu, T. J. and J. S. Tanaka. 1989. "The influence of sample size, estimation method, and model specification on goodness-of-fit assessments in structural equation models." *Journal of Applied Psychology* 74:625-635.

Lance, Charles E. 1989. "Disturbance term regression tests: A note on the computation of standard errors." *Multivariate Behavioral Research* 24:135-141.

Lance, Charles E. 1991. "Evaluation of a structural model relating job satisfaction, organizational commitment, and precursors of voluntary turnover." *Multivariate Behavioral Research* 26:137-162.

Lance, Charles E., John M. Cornwell, and Stanley A. Mulaik. 1988. "Limited information parameter estimates for latent or mixed manifest and latent variable models." *Multivariate Behavioral Research* 23:171-187.

Lavee, Yoav. 1988. "Linear structural relationships (LISREL) in family research." *Journal of Marriage and the Family* 50:937-948.

Lee, Sik-Yum. 1992. "Bayesian analysis of stochastic constraints in structural equation models." *British Journal of Mathematical and Statistical Psychology* 45:93-107.

Lee, Sik-Yum, and Kwan-Moon Leung. 1992. "Estimation of multivariate polychoric and polyserial correlations with missing observations." *British Journal of Mathematical and Statistical Psychology* 45:225-238.

Lee, Sik-Yum, and Wai-Yin Poon. 1992. "Two-level analysis of covariance structures for unbalanced designs with small level-one samples." *British Journal of Mathematical and Statistical Psychology* 45:109-123.

Lee, Sik-Yum, Wai-Yin Poon, and P. M. Bentler. 1990. "A three-stage estimation procedure for structural equation models with polytomous variables." *Psychometrika* 55:45-51.

Lee, Sik-Yum, Wai-Yin Poon, and P. M. Bentler. 1992. "Structural equation models with continuous and polytomous variables." *Psychometrika* 57:89-105.

Lee, Soonmook. 1987. "Model equivalence in covariance structure modeling." Doctoral dissertation. Ohio State University, Columbus, Ohio.

Lee, Soonmook, and Scott Hershberger. 1990. "A simple rule for generating equivalent models in covariance structure modeling." *Multivariate Behavioral Research* 25:313-334.

Levin, Joseph. 1988. "Multiple group factor analysis of multitrait-multimethod matrices." *Multivariate Behavioral Research* 23:469-479.

Loehlin, John C. 1987. *Latent Variable Models: An Introduction to Factor, Path, and Structural Analysis.* Hillsdale, NJ: Erlbaum.

Lomax, Richard G. 1989. "Covariance structure analysis: Extensions and developments." Pp. 171-204 in Bruce Thompson (ed.) *Advances in Social Science Methodology*, Volume 1. Greenwich, CT: JAI Press.

Long, J. Scott, and Pravin K. Trivedi. 1992. "Some specification tests for the linear regression model." *Sociological Methods and Research* 21:161-204.

Long, J. Scott, and Pravin K. Trivedi. 1993. "Some specification tests for the linear regression model." Pp. 66-110 in K. Bollen and R. Stine (eds.) *Testing Structural Equation Models*. Newbury Park: Sage.

Lord, Kenneth R., Myung-Soo Lee, and Paul L. Sauer. 1994. "Program context antecedents of attitude toward radio commercials." *Journal of the Academy of Marketing Science* 22:3-15.

Luijben, Thom C. W. 1991. "Equivalent models in covariance structure analysis." *Psychometrika* 56:653-665.

Luo, Dasen, Stephen A. Petrill, and Lee A. Thompson. 1994. "An exploration of genetic g: Hierarchical factor analysis of cognitive data from the Western Reserve Twin Project." *Intelligence* 18:335-347.

MacCallum, Robert. 1986. "Specification searches in covariance structure modeling." *Psychological Bulletin* 100:107-120.

MacCallum, Robert C. 1990. "The need for alternative measures of fit in covariance structure modeling." *Multivariate Behavioral Research* 25:157-162.

MacCallum, Robert C., and Michael W. Browne. 1993. "The use of causal indicators in covariance structure models: Some practical issues." *Psychological Bulletin* 114:533-541.

MacCallum, Robert C., Mary Roznowski, Corinne M. Mar, and Janet V. Reith. 1994. "Alternative strategies for cross-validation of covariance structure models." *Multivariate Behavioral Research* 29:1-32.

MacCallum, Robert C., Mary Roznowski, and Lawrence B. Necowitz. 1992. "Model modifications in covariance structure analysis: The problem of capitalizing on chance." *Psychological Bulletin* 111:490-504.

Maiti, Sadhan Samar, and Bishwa Nath Mukherjee. 1990. "A note on distributional properties of the Joreskog-Sorbom fit indices." *Psychometrika* 55:721-726.

Maiti, Sadhan Samar, and Bishwa Nath Mukherjee. 1991. "Two new goodness-of-fit indices for covariance matrices with linear structures." *British Journal of Mathematical and Statistical Psychology* 44:153-180.

Marcoulides, George A. 1990. "Evaluation of confirmatory factor analytic and structural equation models using goodness-of-fit indices." *Psychological Reports* 67:669-670.

Marsh, Herbert W. 1989. "Confirmatory factor analysis of multitrait-multimethod data: Many problems and a few solutions." *Applied Psychological Measurement* 13:335-361.

Marsh, Herbert W. 1993. "Stability of individual differences in multiwave panel studies: A comparison of simplex models and one-factor models." *Journal of Educational Measurement* 30:157-183.

Marsh, Herbert W. 1994. "Using the National Longitudinal Study of 1988 to evaluate theoretical models of self-concept: The self-description questionnaire." *Journal of Educational Psychology* 86:439-456.

Marsh, Herbert W., and John R. Balla. 1994. "Goodness of fit in confirmatory factor analysis: The effects of sample size and model parsimony." *Quality and Quantity* 28:185-217.

Marsh, Herbert W., John R. Balla, and Roderick P. McDonald. 1988. "Goodness-of-fit indexes in confirmatory factor analysis: The effect of sample size." *Psychological Bulletin* 103:391-410.

Marsh, Herbert W., and Barbara M. Byrne. 1993. "Confirmatory factor analysis of multitrait-multimethod self-concept data: Between-group and within-group invariance constraints." *Multivariate Behavioral Research* 28:313-349.

Marsh, Herbert W., and David Grayson. 1990. "Public/Catholic differences in the High School and Beyond data: A multigroup structural equation modeling approach to testing mean differences." *Journal of Educational Statistics* 15:199-235.

Marsh, Herbert W., Kit-Tai Hau, Lawrence Roche, Rhonda Craven, John Balla, and Valentina McInerney. 1994. "Problems in the application of structural equation modeling: Comment on Randhawa, Beamer, and Lundberg (1993)." *Journal of Educational Psychology* 86:457-462.

Mason, William M. 1991. "Freedman is right as far as he goes, but there is more, and it's worse: Statisticians could help." Pp. 337-351 in Peter V. Marsden (ed.) *Sociological Methodology 1991*. Washington: American Sociological Association.

Matsueda, Ross L. 1989 "The dynamics of moral beliefs and minor deviance." *Social Forces* 68:428-457.

Matsueda, Ross L. 1992. "Reflected appraisals, parental labeling, and delinquency: Specifying a symbolic interactionist theory." *American Journal of Sociology* 97:1577-1611.

Matsueda, Ross L., and William T. Bielby. 1986. "Statistical power in covariance structure models." Pp. 120-158 in Nancy Brandon Tuma (ed.) *Sociological Methodology 1986.* Washington: American Sociological Association.

Mayer, Lawrence S., and Steven S. Carroll. 1987. "Testing for lagged, cotemporal, and total dependence in cross-lagged panel analysis." *Sociological Methods and Research* 16:187-217.

Mayer, Lawrence S., and Steven S. Carroll. 1988. "Measures of dependence for cross-lagged panel models." *Sociological Methods and Research* 17:93-120.

McArdle, John J. 1994. "Presidential address: Structural factor analysis experiments with incomplete data." *Multivariate Behavioral Research* 29:409-454.

McArdle, J. J., and Raymond B. Cattell. 1994. "Structural equation models of factorial invariance in parallel proportional profiles and oblique confactor problems." *Multivariate Behavioral Research* 29:63-113.

McArdle, J. J., and H. H. Goldsmith. 1990. "Alternative common factor models for multivariate biometric analyses." *Behavior Genetics* 20:569-608.

McArdle, J. J., Fumiaki Hamagami, Merrill F. Elias, and Michael A. Robbins. 1991. "Structural modeling of mixed longitudinal and cross-sectional data." *Experimental Aging Research* 17:29-52.

McArdle, J. Jack, and Roderick P. McDonald. 1984. "Some algebraic properties of the reticular action model for moment structures." *British Journal of Mathematical and Statistical Psychology* 37:234-251.

McCallum, R. Steve. 1990. "Determining the factor structure of the Stanford-Binet Fourth Edition--the right choice." *Journal of Psychoeducational Assessment* 8:436-442.

McDonald, Roderick P. 1981. "The dimensionality of tests and items." *British Journal of Mathematical and Statistical Psychology* 34:100-117.

McDonald, Roderick P. 1994. "The bilevel reticular action model for path analysis with latent variables." *Sociological Methods and Research* 22:399-413.

McDonald, Roderick P., and Harvey Goldstein. 1989. "Balanced versus unbalanced designs for linear structural relations in two-level data." *British Journal of Mathematical and Statistical Psychology* 42:215-232.

McDonald, Roderick P., and Wolfgang M. Hartmann. 1992. "A procedure for obtaining initial values of parameters in the RAM model." *Multivariate Behavioral Research* 27:57-76.

McDonald, Roderick P., and Herbert W. Marsh. 1990. "Choosing a multivariate model: Noncentrality and goodness of fit." *Psychological Bulletin* 107:247-255.

McDonald, Roderick P., Prudence M. Parker, and Tomoichi Ishizuka. 1993. "A scale-invariant treatment for recursive path models." *Psychometrika* 58:431-443.

Millsap, Roger E., and Howard Everson. 1991. "Confirmatory measurement model comparisons using latent means." *Multivariate Behavioral Research* 26:479-497.

Mislevy, Robert J. 1993. "Should 'multiple imputations' be treated as 'multiple indicators'?" *Psychometrika* 58:79-85.

Mislevy, Robert J., and Kathleen M. Sheehan. 1989. "Information matrices in latent-variable models." *Journal of Educational Statistics* 14:335-350.

Mueller, Ralph O., David E. Hutchins, and Daniel E. Vogler. 1990. "Validity and reliability of the Hutchins Behavior Inventory: A confirmatory maximum likelihood analysis." *Measurement and Evaluation in Counseling and Development* 22:203-214.

Mulaik, Stanley A. 1988. "Confirmatory factor analysis." Pp. 259-288 in John Nesselroade and Raymond Cattell (eds.) *Handbook of Multivariate Experimental Psychology* 2nd ed. New York: Plenum.

Mulaik, Stanley A., Larry R. James, Judith Van Alstine, Nathan Bennett, Sherri Lind, and C. Dean Stilwell. 1989. "Evaluation of goodness-of-fit indices for structural equation models." *Psychological Bulletin* 105:430-445.

Muthen, Bengt. 1984. "A general structural equation model with dichotomous, ordered categorical, and continuous latent variable indicators." *Psychometrika* 49:115-132.

Muthen, Bengt. 1987. "Tobit factor analysis." *British Journal of Mathematical and Statistical Psychology* 42:241-250.

Muthen, Bengt. 1989a. "Multiple-group structural modelling with non-normal continuous variables." *British Journal of Mathematical and Statistical Psychology* 42:55-62.

Muthen, Bengt. 1989b. "Latent variable modeling in heterogeneous populations." *Psychometrika* 54:557-585.

Muthen, Bengt O. 1993. "Goodness of fit with categorical and other nonnormal variables." Pp. 205-234 in K. Bollen and R. Stine (eds.) *Testing Structural Equation Models*. Newbury Park: Sage.

Muthen, Bengt. 1994. "Multilevel covariance structure analysis." *Sociological Methods and Research* 22:376-398.

Muthen, Bengt, and David Kaplan. 1992. "A comparison of some methodologies for the factor analysis of non-normal Likert variables: A note on the size of the model." *British Journal of Mathematical and Statistical Psychology* 45:19-30.

Muthen, Bengt, David Kaplan, and Michael Hollis. 1987. "On structural equation modeling with data that are not missing completely at random." *Psychometrika* 52:431-462.

Neale, M. C. 1993. *Mx Structural Equation Modeling*. Richmond, VA: Medical College of Virginia.

Newcomb, Michael D., and P. M. Bentler. 1987. "Self-report methods of assessing health status and health service utilization: A hierarchical confirmatory analysis." *Multivariate Behavioral Research* 22:415-436.

Norris, Fran H., and Krzysztof Kaniasty. 1992. "A longitudinal study of the effects of various crime prevention strategies on criminal victimization, fear of crime, and psychological distress." *American Journal of Community Psychology* 20:625-648.

Nunnally, Jum C., and Ira H. Bernstein. 1994. *Psychometric Theory*. 3rd ed. New York: McGraw-Hill.

O'Brien, Robert M., and Pamela Homer. 1987. "Corrections for coarsely categorized measures: LISREL's polyserial and polychoric correlations." *Quality and Quantity* 21:349-360.

Offodile, O. Felix, Kingsley O. Ugwu, and Leslie A. Hayduk. 1993. "Analysis of the causal structures linking process variables to robot repeatability and accuracy." *Technometrics* 35:421-435.

Ogasawara, Haruhiko. 1990. "Covariance structure model when the factor means and the covariances are functions of the third variable." *Japanese Psychological Research* 32:19-25.

O'Grady, Kevin E., and Deborah R. Medoff. 1991. "Rater reliability: A maximum likelihood confirmatory factor-analytic approach." *Multivariate Behavioral Research* 26:363-387.

Opwis, K., A. Gold, and W. Schneider. 1987. "Moglichkeiten der kreuzvalidierung von strukturgleichungsmodellen." *Psychologische Beitrage* 29:60-77 (with English abstract).

Oud, Johan H., John H. van den Bercken, and Raymond J. Essers. 1990. "Longitudinal factor score estimation using the Kalman filter." *Applied Psychological Measurement* 14:395-418.

Owens, Timothy J. 1993. "Accentuate the positive--and the negative: Rethinking the use of self-esteem, self-deprecation, and self-confidence." *Social Psychology Quarterly* 56:288-299.

Palmquist, Bradley, and Donald P. Green. 1992. "Estimation of models with correlated measurement errors from panel data." Pp. 119-146 in Peter V. Marsden (ed.) *Sociological Methodology 1992*. Washington: American Sociological Association.

Pedersen, Nancy L., Robert Plomin, and G. E. McClearn. 1994. "Is there G beyond g? (Is there genetic influence on specific cognitive abilities independent of genetic influence on general cognitive abilities?)" *Intelligence* 18:133-143.

Pfeifer, Andreas, and Peter Schmidt. 1987. *Die Analyse Komplexer Strukturgleichungsmodelle*. Stuttgart: Gustav Fischer Verlag.

Poon, Wai-Yin, and Sik-Yum Lee. 1987. "Maximum likelihood estimation of multivariate polyserial and polychoric correlation coefficients." *Psychometrika* 52:409-430, with errata from 53:301.

Poon, Wai-Yin, and Sik-Yum Lee. 1992. "Statistical analysis of continuous and polytomous variables in several populations." *British Journal of Mathematical and Statistical Psychology* 45:139-149.

Potthast, Margaret J. 1993. "Confirmatory factor analysis of ordered categorical variables with large models." *British Journal of Mathematical and Statistical Psychology* 46:273-286.

Powers, Daniel A. 1993. "Endogenous switching regression models with limited dependent variables." *Sociological Methods and Research* 22:248-273.

Raffalovich, Lawrence E., and George W. Bohrnstedt. 1987. "Common, specific, and error variance components of factor models: Estimation with longitudinal data." *Sociological Methods and Research* 15:385-405.

Raftery, Adrian E. 1993. "Bayesian model selection in structural equation models." Pp. 163-180 in K. Bollen and R. Stine (eds.) *Testing Structural Equation Models*. Newbury Park: Sage.

Ratner, Pamela A., Joan L. Bottorff, Joy L. Johnson, and Leslie A. Hayduk. 1994. "The interaction effects of gender within the health promotion model." *Research in Nursing and Health* 17:341-350.

Reuterberg, Sven-Eric, and Jan-Eric Gustafsson. 1992. "Confirmatory factor analysis and reliability: Testing measurement model assumptions." *Educational and Psychological Measurement* 52:795-811.

Reynolds, Arthur J., and Herbert J. Walberg. 1991. "A structural model of science achievement." *Journal of Educational Psychology* 83:97-107.

Reynolds, Arthur J., and Herbert J. Walberg. 1992. "A structural model of high school mathematics outcomes." *Journal of Educational Research* 85:150-158.

Rigdon, Edward E. forthcoming. "A necessary and sufficient identification rule for structural models." *Multivariate Behavioral Research*.

Rindskopf, David, and Tedd Rose. 1988. "Some theory and applications of confirmatory second-order factor analysis." *Multivariate Behavioral Research* 23:51-67.

Romney, David M. 1987. "A simplex model of the paranoid process: Implications for diagnosis and prognosis." *Acta Psychiatrica Scandinavica* 75:651-655.

Rosenberg, Morris, Carmi Schooler, and Carrie Schoenbach. 1989. "Self-esteem and adolescent problems: Modeling reciprocal effects." *American Sociological Review* 54:1004-1018.

Rowe, David C., Alexander T. Vazsonyi, and Daniel J. Flannery. 1994. "No more than skin deep: Ethnic and racial similarity in development process." *Psychological Review* 101:396-413.

Sachs, John. 1992. "Covariance structure analysis of a test of moral orientation and moral judgment." *Educational and Psychological Measurement* 52:825-833.

Saris, Willem E. 1988. "A measurement model for psychophysical scaling." *Quality and Quantity* 22:417-433.

Saris, W. E., J. Den Ronden, and A. Satorra. 1987. "Testing structural equation models." Pp. 202-220 in Peter Cuttance and Russell Ecob (eds.) *Structural Modeling by Example*. New York: Cambridge University Press.

Saris, Willem E., and Harm Hartman. 1990. "Common factors can always be found but can they also be rejected?" *Quality and Quantity* 24:471-490.

Saris, Willem E., and Albert Satorra. 1993. "Power evaluations in structural equation models." Pp. 181-204 in K. Bollen and R. Stine (eds.) *Testing Structural Equation Models*. Newbury Park: Sage.

Saris, Willem E., Albert Satorra, and Dag Sorbom. 1987. "The detection and correction of specification errors in structural equation models." Pp. 105-129 in Clifford C. Clogg (ed.) *Sociological Methodology 1987*. Washington: American Sociological Association.

Saris, Willem E., and Bas Van Den Putte. 1988. "True scores or factor models: A secondary analysis of the ALLBUS-test-retest data." *Sociological Methods and Research* 17:123-157.

Satorra, Albert. 1989. "Alternative test criteria in covariance structure analysis: A unified approach." *Psychometrika* 54:131-151.

Satorra, Albert. 1990. "Robustness issues in structural equation modeling: A review of recent developments." *Quality and Quantity* 24:367-386.

Satorra, Albert. 1992. "Asymptotic robust inferences in the analysis of mean and covariance structures." Pp. 249-279 in Peter V. Marsden (ed.) *Sociological Methodology 1992*. Washington: American Sociological Association.

Scanzoni, John, and Deborah D. Godwin. 1990. "Negotiation effectiveness and acceptable outcomes." *Social Psychology Quarterly* 53:239-251.

Scheines, Richard, and Peter Spirtes. 1992. "Finding latent variable models in large data bases." *International Journal of Intelligent Systems* 7:609-621.

Scheines, Richard, Peter Spirtes, Clark Glymour, and Christopher Meek. 1994. *Tetrad II: Tools for Causal Modeling, User's Manual.* Erlbaum Statistical Software.

Schoenberg, Ronald. 1989. "Covariance structure models." *Annual Review of Sociology* 15:425-440.

Schoenberg, Ronald, and Gerhard Arminger. 1988. "LINCS: Linear covariance structure analysis." *Multivariate Behavioral Research* 23:271-273.

Schriesheim, Chester A., Esther Solomon, and Richard E. Kopelman. 1989. "Grouped versus randomized format: An investigation of scale convergent and discriminant validity using LISREL confirmatory factor analysis." *Applied Psychological Measurement* 13:19-32.

Seidel, Gerhard, and Cornelia Eicheler. 1990. "Identification structure of linear structural models." *Quality and Quantity* 24:345-365.

Sheehan, T. Joseph. 1987. "The importance of knowing when to stop considering alternative structural equation models." *Evaluation and the Health Professions* 10:101-105.

Sibrian, Ricardo, and Robert Elston. 1990. "Reciprocal causal influences among malnutrition, growth retardation, and diarrhea in preschool children." *American Journal of Human Biology* 2:235-243.

Sikkel, Dirk, and Gerard Jelierse. 1987. "Retrospective questions and correlations." *Psychometrika* 52:251-261.

Smith, Leigh M. 1989. "Is it fitting? Comments on the LISREL analysis by Stoner and Arora of variables affecting the psychological health of strikers." *Journal of Occupational Psychology* 62:257-262.

Sobel, Michael E. 1986. "Some new results on indirect effects and their standard errors in covariance structure models." Pp. 159-186 in Nancy Brandon Tuma (ed.) *Sociological Methodology 1986.* Washington: American Sociological Association.

Sobel, Michael E. 1987. "Direct and indirect effects in linear structural equation models." *Sociological Methods and Research* 16:155-176.

Sobel, Michael E. 1990. "Effect analysis and causation in linear structural equation models." *Psychometrika* 55:495-515.

Sorbom, Dag. 1989. "Model modification." *Psychometrika* 54:371-383.

Spirtes, Peter. 1991. "Building causal graphs from statistical data in the presence of latent variables." Manuscript. Department of Philosophy, Carnegie-Mellon University, Pittsburgh, Pennsylvania.

Spirtes, Peter, and Clark Glymour. 1991. "An algorithm for fast recovery of sparse causal graphs." *Social Science and Computer Review* 9:62-72.

Spirtes, Peter, Clark Glymour and Richard Scheines. 1991a. "From probability to causality." *Philosophical Studies* 64:1-36.

Spirtes, Peter, Clark Glymour, and Richard Scheines. 1991b. "Causal hypotheses, statistical inference and automated model specification." Manuscript. Department of Philosophy, Carnegie-Mellon University, Pittsburgh, Pennsylvania.

Spirtes, Peter, Clark Glymour, and Richard Scheines. 1993. *Causation, Prediction, and Search*. New York: Springer-Verlag.

Spirtes, P., C. Glymour, R. Scheines, C. Meek, S. Fienberg, and E. Slate. 1991. "Prediction and experimental design with graphical causal models." Manuscript. Department of Philosophy, Carnegie-Mellon University, Pittsburgh, Pennsylvania.

Spirtes, Peter, Richard Scheines, and Clark Glymour. 1990a. "Simulation studies of the reliability of computer-aided model specification using the TETRAD II, EQS, and LISREL programs." *Sociological Methods and Research* 19:3-66.

Spirtes, Peter, Richard Scheines, and Clark Glymour. 1990b. "Reply to Comments." *Sociological Methods and Research* 19:107-121.

SPSS Inc. 1993. *SPSS LISREL 7 and PRELIS*. Chicago: SPSS Inc.

Stack, Steve, Ira Wasserman, and Augustine Kposowa. 1994. "The effects of religion and feminism on suicide ideology: An analysis of national survey data." *Journal for the Scientific Study of Religion* 33:110-121.

Steiger, James H. 1988. "Aspects of person-machine communication in structural modeling of correlations and covariances." *Multivariate Behavioral Research* 23:281-290.

Steiger, James H. 1989. *EzPath Causal Modeling: A Supplementary Module for SYSTAT and SYGRAPH*. Evanston, IL: SYSTAT Inc.

Steiger, James H. 1990. "Structural model evaluation and modification: An interval estimation approach." *Multivariate Behavioral Research* 25:173-180.

Steiger, James H., Alexander Shapiro, and Michael W. Browne. 1985. "On the multivariate asymptotic distribution of sequential chi-square statistics." *Psychometrika* 50:253-264.

Stelzl, Ingeborg. 1986. "Changing a causal hypothesis without changing the fit: Some rules for generating equivalent path models." *Multivariate Behavioral Research* 21:309-331.

Stelzl, Ingeborg. 1991. "Rival hypotheses in linear structure modeling: Factor rotation in confirmatory factor analysis and latent path analysis." *Multivariate Behavioral Research* 26:199-225.

Steyer, Rolf, and Manfred J. Schmitt. 1990. "Latent state-trait models in attitude research." *Quality and Quantity* 24:427-445.

Stone, Clement A., and Michael E. Sobel. 1990. "The robustness of estimates of total indirect effects in covariance structure models estimated by maximum likelihood." *Psychometrika* 55:337-352.

Stryker, Sheldon, and Richard T. Serpe. 1994. "Identity salience and psychological centrality: Equivalent, overlapping, or complementary concepts?" *Social Psychology Quarterly* 57:16-35.

Suyapa, E., M. Silvia, and Robert C. MacCallum. 1988. "Some factors affecting the success of specification searches in covariance structure modeling." *Multivariate Behavioral Research* 23:297-326.

Swaminathan, H. 1991. "Analysis of covariance structures." Pp. 97-124 in R. K. Hambleton and J. N. Zaal (eds.) *Advances in Educational and Psychological Testing*. Boston: Kluwer Academic.

Tanaka, J. S. 1990. "Towards the second generation of structural modeling." *Multivariate Behavioral Research* 25:187-191.

Tanaka, J. S. 1993. "Multifaceted conceptions of fit in structural equation models." Pp. 10-39 in K. Bollen and R. Stine (eds.) *Testing Structural Equation Models*. Newbury Park: Sage.

Tanaka, J. S., and G. J. Huba. 1989. "A general coefficient of determination for covariance structure models under arbitrary GLS estimation." *British Journal of Mathematical and Statistical Psychology* 42:233-239.

Thompson, B., and G. M. Borrello. 1992. "Measuring second-order factors using confirmatory methods: An illustration with the Hendrick-Hendrick Love Instrument." *Educational and Psychological Measurement* 52:69-77.

Thon, Rali. 1993. "Staff organization and pupil outcomes in multi-ethnic junior high schools." Doctoral dissertation, Tel Aviv University, Tel Aviv, Israel.

Thornberry, Terence, Alan Lizotte, Marvin Krohn, Margaret Farnworth, and Sung Joon Jang. 1994. "Delinquent peers, beliefs, and delinquent behavior: A longitudinal test of interactional theory." *Criminology* 32:47-83.

Turner, Stephen P. 1987a. "Cause, concepts, measures and the underdetermination of theory by data." *International Review of Sociology* 3:249-271.

Turner, Stephen P. 1987b. "Underdetermination and the promise of statistical sociology." *Sociological Theory* 5:172-184.

Undheim, Johan Olav, and Jan-Eric Gustafsson. 1987. "The hierarchical organization of cognitive abilities: Restoring general intelligence through the use of linear structural relations (LISREL)." *Multivariate Behavioral Research* 22:149-171.

Verma, T. S., and Judea Pearl. 1990. "Equivalence and synthesis of causal models." Pp. 220-227 in *Proceedings of the Sixth Conference on Uncertainty in Artificial Intelligence*. July 27-29, Cambridge Massachusetts.

Vittadini, Giorgio. 1989. "Indeterminacy problems in the LISREL model." *Multivariate Behavioral Research* 24:397-414.

Von Eye, Alexander, and Clifford C. Clogg (eds.). 1994. *Latent Variables Analysis: Applications for Developmental Research*. Thousand Oaks, CA: Sage.

Waller, Niels G. 1993. "Seven confirmatory factor analysis programs: EQS, EzPATH, LINCS, LISCOMP, LISREL 7, SIMPLIS, and CALIS." *Applied Psychological Measurement* 17:73-100.

Watson, Betty U., and Theodore K. Miller. 1993. "Auditory perception, phonological processing, and reading ability/disability." *Journal of Speech and Hearing Research* 36:850-863.

Westholm, Anders. 1987. "Measurement error in causal analysis of panel data: Attenuated versus inflated relationships." *Quality and Quantity* 21:3-20.

Wheaton, Blair. 1987. "Assessment of fit in overidentified models with latent variables." *Sociological Methods and Research* 16:118-154.

Wheaton, Blair, Bengt Muthen, Duane F. Alwin, and Gene F. Summers. 1977. "Assessing reliability and stability in panel models." Pp. 84-136 in David Heise (ed.) *Sociological Methodology 1977*. San Francisco: Jossey-Bass.

Whitbeck, Les B., Dan R. Hoyt, Ronald L. Simons, Rand D. Conger, Glen H. Elder Jr., Frederick O. Lorenz, and Shirley Huck. 1992. "Intergenerational continuity of parental rejection and depressed affect." *Journal of Personality and Social Psychology* 63:1036-1045.

Willett, John B., and Aline G. Sayer. 1994. "Using covariance structure analysis to detect correlates and predictors of individual change over time." *Psychological Bulletin* 116:363-381.

Wiseman, Richard L., Mitchell R. Hammer, and Hiroko Nishida. 1989. "Predictors of intercultural communication competence." *International Journal of Intercultural Relations* 13:349-370.

Wood, Phillip, and David Brown. 1994. "The study of intraindividual differences by means of dynamic factor models: Rationale, implementation, and interpretation." *Psychological Bulletin* 116:166-186.

Wothke, Werner. 1993. "Nonpositive definite matrices in structural modeling." Pp. 256-293 in K. Bollen and R. Stine (eds.) *Testing Structural Equation Models*. Newbury Park: Sage.

Wothke, Werner, and Michael W. Browne. 1990. "The direct product model for the MTMM matrix parameterized as a second order factor analysis model." *Psychometrika* 55:255-262.

Xie, Yu. 1989. "Structural equation models for ordinal variables: An analysis of occupational destination." *Sociological Methods and Research* 17:325-352.

Yung, Yiu-Fai, and Peter M. Bentler. 1994. "Bootstrap-corrected ADF test statistics in covariance structure analysis." *British Journal of Mathematical and Statistical Psychology* 47:63-84.

Zeller, Richard A. 1987. "Comment on common, specific, and error variance components of factor models: Estimation with longitudinal data." *Sociological Methods and Research* 15:406-419.

Zeller, Richard A., and Edward G. Carmines. 1980. *Measurement in the Social Sciences: The Link between Theory and Data*. New York: Cambridge University Press.

Index